行业特色翻译系列教材

武汉科技大学2022年度研究生教育质量工程教材建设项目
资助出版（文件号：武科大研发【2022】59号）

冶金科技英语翻译教程

主　编　姚　刚　朱明炬
副主编　方庆华　徐　敏　冯　婕　韩　静

武汉大学出版社

图书在版编目(CIP)数据

冶金科技英语翻译教程/姚刚,朱明炬主编. -- 武汉：武汉大学出版社,2025.5. -- 行业特色翻译系列教材/熊伟总主编. -- ISBN 978-7-307-24684-3

Ⅰ.TF

中国国家版本馆 CIP 数据核字第 2024DC4284 号

责任编辑：邓 喆　　责任校对：汪欣怡　　版式设计：马 佳

出版发行：**武汉大学出版社**　（430072　武昌　珞珈山）
　　　　　（电子邮箱：cbs22@whu.edu.cn　网址：www.wdp.com.cn）
印刷：湖北金海印务有限公司
开本：787×1092　1/16　印张：17.5　字数：367 千字　插页：1
版次：2025 年 5 月第 1 版　　2025 年 5 月第 1 次印刷
ISBN 978-7-307-24684-3　　定价：69.00 元

版权所有，不得翻印；凡购买我社的图书，如有质量问题，请与当地图书销售部门联系调换。

前　言

冶金，是"冶炼金属"的意思，指从金属矿物中获取金属或金属化合物的过程，分为化学冶金和物理冶金两大类。化学冶金通过化学反应从矿石中提取金属或金属化合物，物理冶金则通过成型、加工等方法制备有一定性能的金属或合金材料。

冶金工业涵盖勘探、开采、精选、烧结金属矿石并对其进行冶炼、提取、加工成金属和金属材料的所有工业行业，是国民经济的基础工业，在国民经济发展中起着不可替代的作用。很难想象，没有钢铁、铜、锌、锡、铝等金属，这个世界会是什么样！

冶金工业每年创造的产值也十分惊人。根据国家统计局数据，2022年我国累计生产生铁86383万吨、粗钢101300万吨、钢材134034万吨、十种有色金属6774.3万吨。在行业利润方面，2022年全年，我国金属矿业行业累计营业收入172057.7亿元，利润4275.4亿元，占全国规模以上工业企业利润总额(84038.5亿元)的近5.1%。[①]

冶金工业的发展离不开大规模的进出口活动。根据中国海关总署的统计，2022年我国累计进口铁矿砂及其精矿110686万吨、铜矿砂及其精矿2527.1万吨、铝矿砂及其精矿12547万吨。与此同时，我国共出口钢材6732.3吨、未锻造铜及制成品91.65万吨、铝材660.36万吨。[②]

冶金工业的发展同样离不开人员和技术的对外交流。熟悉冶金行业发展的人士都知道：鞍钢、包钢、武钢等钢铁公司的建成有着苏联专家的巨大贡献；宝钢的建设，日本提供了技术支持与设备……随着"一带一路"倡议的实施，越来越多的中国冶金企业正走向世界：河钢集团收购塞尔维亚斯梅代雷沃钢厂，敬业集团收购英国钢铁公司，青山集团在印尼投资建成大型镍铁项目等，标志着我国冶金企业已经实现了从"引进来"到"走出去"的历史性转变。

冶金原料和产品的进出口、冶金人员和技术的对外交流都离不开翻译，这类翻译可称为"冶金翻译"。冶金翻译很重要，因为如果没有冶金翻译，冶金贸易、投融资和冶金人员、技术交流等就无从谈起，势必会严重影响国民经济的发展。但遗憾的是，根据我们搜

---

① 国家统计局. 2022年冶金工业生产与利润数据[EB/OL].(2022-12-31)[2024-10-20]. http://www.stats.gov.cn/

② 中国海关总署. 2022年冶金工业进出口数据[EB/OL].(2022-12-31)[2025-03-20]. http://www.customs.gov.cn/

集到的数据，迄今为止唯一一本讨论冶金翻译的书只有吴小力老师编著的《冶金科技英语口译教程》，该书由冶金工业出版社于 2013 年出版。

为了填补冶金笔译教材的空白，我们编写了这本教程。教程的出版获得"武汉科技大学 2022 年度研究生教育质量工程教材建设项目"资助。本教程主要面向英语或翻译专业各层次学生，不仅传授冶金知识，同时讲解翻译技巧，理论与实践紧密结合。它既可作为英语或翻译专业"冶金翻译"或"冶金科技翻译"课程教学用书，也可供冶金专业学生学习专业英语及冶金翻译使用。

本教程有以下五个主要特点：

1. 教程以黑色冶金为主，辅以少量有色冶金内容，涵盖了黑色冶金从原料到炼焦、炼铁、炼钢、轧制等整个流程。内容全面，没有任何冶金基础的学生也能通过教程系统地获得必要的冶金知识，具备一定的冶金翻译能力。

2. 教程包含的译例丰富，不仅每篇课文都可以当作译例，"分析与讲解"及"专题讨论"部分还提供了大量的译例，方便老师根据教学时间和学生的水平选择合适的译例，也方便学生从实践中学习。

3. 教程所选材料由短至长、由浅入深，难度循序渐进，符合教学规律，不会让学生产生畏难情绪，同时也便于部分读者将其当作自学用书。

4. 教程中给出的参考译文大多出自编者之手，对教材原文的分析与讲解基于编者的亲身翻译实践，有感而发，真实生动，容易引起师生的共鸣。

5. 教程以英译汉为主，兼顾汉译英。考虑到英译汉的难度小于汉译英，所以英译汉以篇章为主，汉译英以句子为主。

本教程可供一学期的教学使用，总计需要 32 至 40 学时。根据教学学时数的不同和学生水平的高低，教师可指定整篇课文，或选择课文中某些段落、句子让学生在课前进行翻译。授课时，建议以讨论为主，多让学生分析，教师以启发式点评为主。

本教程是集体心血的结晶，具体分工如下：

姚刚：第四课和第五课

朱明炬：前言；第一课和第八课；附录一、附录二

方庆华：第六课和第七课

徐敏：第二课和第三课

冯婕：第九课

韩静：第十课

附录三由每课负责老师集体完成。

全书的统稿、通审、校对等工作，由姚刚和朱明炬完成。

我们本都是英语语言文学或英语教育专业出身，对于冶金所知有限。之所以决定编写本教程，一是因为武汉科技大学原为冶金类高校，钢铁冶金是我校的优势学科，作为学校

的一分子，我们有责任将钢铁冶金发扬光大；二是因为冶金翻译虽小众，但存在切切实实的现实需求，冶金翻译人才培养关乎冶金工业的长远发展，关乎国民经济的发展大计，而冶金翻译教材的编写则是冶金翻译人才培养的关键。因此，虽然深知自身学识有限，但我们仍然鼓起勇气完成了教程的编写工作。可以想见，教程难免存在不足之处，衷心希望教程使用者不吝赐教，我们在此表示衷心的感谢！

编者

2024 年 10 月

目 录

第一课 概论 ... 1
一、金属和冶金 ... 1
二、冶金英语 ... 3
三、冶金科技英语翻译 ... 13

第二课 铁 ... 28
一、科技术语 ... 28
二、英语原文 ... 30
三、分析与讲解 ... 35
四、参考译文 ... 38
五、专题讨论 ... 42

第三课 磁铁矿 ... 48
一、科技术语 ... 48
二、英语原文 ... 50
三、分析与讲解 ... 52
四、参考译文 ... 55
五、专题讨论 ... 57

第四课 高炉炼铁用焦炭的生产 ... 64
一、科技术语 ... 64
二、英语原文 ... 65
三、分析与讲解 ... 69
四、参考译文 ... 73
五、专题讨论 ... 75

第五课　高炉是如何运转的 …… 82
一、科技术语 …… 82
二、英语原文 …… 83
三、分析与讲解 …… 86
四、参考译文 …… 89
五、专题讨论 …… 94

第六课　直接还原工艺 …… 100
一、科技术语 …… 100
二、英语原文 …… 101
三、分析与讲解 …… 104
四、参考译文 …… 110
五、专题讨论 …… 112

第七课　碱性氧气转炉炼钢工艺 …… 118
一、科技术语 …… 118
二、英语原文 …… 119
三、分析与讲解 …… 129
四、参考译文 …… 136
五、专题讨论 …… 144

第八课　电弧炉 …… 152
一、科技术语 …… 152
二、英语原文 …… 153
三、分析与讲解 …… 162
四、参考译文 …… 164
五、专题讨论 …… 171

第九课　轧制（金属加工） …… 179
一、科技术语 …… 179
二、英语原文 …… 180
三、分析与讲解 …… 186

四、参考译文 ·· 190
　　五、专题讨论 ·· 194

第十课　有色金属：生产和历史 ·· 199
　　一、科技术语 ·· 199
　　二、英语原文 ·· 205
　　三、分析与讲解 ·· 218
　　四、参考译文 ·· 226
　　五、专题讨论 ·· 234

附录一　汉英翻译练习 ·· 241

附录二　汉英翻译练习参考答案 ·· 250

附录三　课后练习参考答案 ··· 263

参考文献 ·· 270

第一课 概 论

> Translating consists in reproducing in the receptor language the closest natural equivalent of the source language message, first in terms of meaning and secondly in terms of style.
>
> ——By Eugene A. Nida[①]

一、金属和冶金

金属(metal)指"any of a class of substances characterized by high electrical and thermal conductivity as well as by malleability, ductility, and high reflectivity of light"[②](容易导电导热、富有延展性、对可见光强烈反射的一类物质)。除汞(hydrargyrum)以外,金属在常温下都是固体的单质或化合物。绝大多数金属在自然界中以化合态存在,只有金(gold)、银(silver)、铂(platinum)、铋(bismuth)以单质的游离态存在。在已确认的100多种元素中,金属元素有90多种。

铁(iron)、铬(chromium)、锰(manganese)及这三种元素的合金称为黑色金属(ferrous metal),一般主要指铁及其合金,如生铁、钢、铁合金等。铁、铬、锰以外的所有金属及其合金称为有色金属(non-ferrous metal)。1958年我国将64种金属列为有色金属,其中又有重金属(heavy metal)、轻金属(light metal)、贵金属(precious metal/noble metal)、稀有金属(rare metal)等之分。这64种有色金属包括:铝(aluminum)、镁(magnesium)、钾(potassium)、钠(sodium)、钙(calcium)、锶(strontium)、钡(barium)、钒(vanadium)、铜(copper)、铅(lead)、锌(zinc)、锡(tin)、钴(cobalt)、镍(nickel)、锑(antimony)、汞(mercury)、镉(cadmium)、铋(bismuth)、金(gold)、银(silver)、铂(platinum)、钌(ruthenium)、铑(rhodium)、钯(palladium)、锇(osmium)、铱(iridium)、铍(beryllium)、

[①] Eugene A. Nida & Charles R. Taber. *The Theory and Practice of Translation*. 上海:上海外语教育出版社,2004:12.

[②] *Britannica Encyclopedia*. Metal [EB/OL]. https://www.britannica.com/search?query=Metal.

锂(lithium)、铷(rubidium)、铯(cesium)、钛(titanium)、锆(zirconium)、铪(hafnium)、铌(niobium)、钽(tantalum)、钨(tungsten)、钼(molybdenum)、镓(gallium)、铟(indium)、铊(thallium)、锗(germanium)、铼(rhenium)、镧(lanthanum)、铈(cerium)、镨(praseodymium)、钕(neodymium)、钐(samarium)、铕(europium)、钆(gadolinium)、铽(terbium)、镝(dysprosium)、钬(holmium)、铒(erbium)、铥(thulium)、镱(ytterbium)、镥(lutetium)、钪(scandium)、钇(yttrium)、硅(silicon)、硼(boron)、硒(selenium)、碲(tellurium)、砷(arsenic)、钍(thorium)。其中，金、银、铜、汞、铅、镉等为重金属；铝、镁、钙、钛、钾、锶、钡等为轻金属；金、银、铂、钌、铑、钯、锇、铱为贵金属；锂、铍、钛、钒、锗、铌、钼、铯、镧、钨、钍等为稀有金属。

金属的使用是人类文明史上的重大事件。"人类和自然斗争的历史大致可分为两大时代——石器时代和金属时代。而金属时代又分为铜器时代和铁器时代，它标志着人类生产大发展的三个飞跃阶段，也是记载人类文化进展的三个里程碑。"[1]

金属之所以得到广泛应用，是由于金属有许多优良的性能(properties)。金属材料的性能包括使用性能(service performance)和工艺性能(processing properties)。前者包括物理性能(physical properties)、化学性能(chemical properties)、力学性能(mechanical properties)，后者包括铸造性能(castability)、锻造性能(forgeability)、焊接性能(weldability)及切削加工性能(machinability)。

金属的物理性能包括密度(density)、熔点(melting point)、导热性(thermal/heat conductivity)、导电性(electrical conductivity/electroconductivity)、热膨胀性(thermal expansibility)、磁性(magnetic property/performance)等。化学性能包括耐腐蚀性(corrosion resistance)、抗氧化性(oxidation resistance)、化学稳定性(chemical stability)等。力学性能包括强度(strength)、塑性(ductility/plasticity)、硬度(hardness)、韧性(toughness)、弯曲性能(bending property)、疲劳强度(fatigue strength)等。

要想得到金属及金属制成品，冶金必不可少。作为一个词语，冶金有广义、狭义之别。广义的冶金包括金属矿物(metallic minerals)的勘探(prospecting)、开采(mining)、选矿(beneficiation)、冶炼(smelting)，以及有色金属及其合金、化合物的加工等过程。狭义的冶金是指矿石或精矿(concentrate)的冶炼。由于科学技术的进步和工业的发展，采矿、选矿和金属加工已各自形成独立的学科[2]，所以我们今天所说的冶金，大多指狭义的冶金，也就是"提取冶金"(extractive metallurgy)。

提取冶金的方法主要分为三大类：火法冶金(pyrometallurgy)、湿法冶金(hydrometallurgy)和电冶金(electrometallurgy)。火法冶金指矿石在高温下发生一系列物理和化学变化，使其

---

[1] 艾星辉. 金属学[M]. 北京：冶金工业出版社，2009.
[2] 周兰花. 冶金原理[M]. 重庆：重庆大学出版社，2016.

中的金属和杂质(impurities)分离，获得较纯金属的过程。湿法冶金指在低温下(一般低于100℃，现代湿法冶金开发的高温高压过程温度可达230℃至300℃)使用熔剂(flux)处理矿石或精矿，使所要提取的金属溶解于溶液中，而其他杂质不溶解，然后通过液固分离(liquid-solid separation)制得含金属的净化液，最后再从净化液中将金属提取和分离出来。电冶金利用电能产生高温来提取、精炼金属，又分电热冶金(electrothermal metallurgy)和电化学冶金(electrochemical metallurgy)两类。前者通过将电能转化为热能，在高温下提炼金属；后者利用电化学反应，使金属从含金属盐的溶液或熔体中析出。①

根据提炼出来的金属种类，冶金还可以分为黑色冶金(ferrous metallurgy)和有色冶金(non-ferrous metallurgy)两大类。顾名思义，前者提炼出来的是黑色金属，后者提炼出来的是有色金属。通常所说的黑色冶金，主要指钢铁冶金，即炼铁、炼钢。

冶金是人类生产实践中最重要的活动之一。青铜、铁、钢等金属的先后发现、发明和使用，彻底改变了人类社会的面貌，使得工业化时代的到来成为可能。

我国是世界上最大的发展中国家，也是冶金大国。2021年，全球钢铁产量19.5亿吨，我国就达到了10.3亿吨；全世界最大的10家钢铁企业中，我国占了6家；我国铝产量位居世界第一，约占世界总产量的60%；全球最大的锌生产国仍然是我国，2021年我国锌产量约为420万吨，远远超过排名第二的秘鲁(160万吨)和排名第三的澳大利亚(130万吨)。

尽管如此，我国尚未成为冶金强国。尽管我国生产了全世界近53%的钢铁，但2021年我国仍然进口了1427万吨钢材，这主要是由于技术相对落后，有些种类的钢材我国还无法生产，只能依赖进口。正因为此，我国的冶金国际贸易、冶金技术的对外交流还需要不断加强，冶金领域的英语学习和应用的重要性由此凸显。

二、冶金英语

所谓"冶金英语"，就是以冶金为话题(topic)的英语表达。

虽然冶金产品与我们的日常生活紧密相关，但以冶金为话题的英语语篇(discourse②)在日常生活中却并不常见，主要见于冶金原料及产品贸易、冶金市场报道以及冶金科技交流活动中。服务于冶金原料及产品贸易的英语，可称为冶金贸易英语；服务于冶金市场报道的英语，可称为冶金新闻英语；服务于冶金科技交流的英语，可称为冶金科技英语。

冶金贸易英语除了包含一些冶金专业术语之外，与其他种类的贸易英语并无太大区别，其文体上表现出来的是贸易英语的特点。例如下面这份铁矿细粉买卖合同：

---

① 周兰花. 冶金原理[M]. 重庆：重庆大学出版社，2016.
② 语言学中，语篇(discourse)这一术语有广义、狭义之分，广义的语篇包括口头话语(oral discourse)和书面文本(written text)，狭义的语篇只指书面的文本。这里使用的是广义上的语篇。

SALE AND PURCHASE CONTRACT OF IRON ORE FINES

Contract No.: 11206/A

Date: July 26, 2022

Seller: Central China Iron and Steel Company

Address: 2222 Heping Avenue, Wuhan, China

Telephone: +86 27 88888888

Fax: +86 27 88888888

Represented by: Ma Dawei

AND

Buyer:

Address:

Telephone:

Fax:

Email:

This contract is made by and between the Buyer and the Seller whereby the Buyer agrees to buy and the Seller agrees to sell the under mentioned goods on the terms and conditions stated below:

CLAUSE 1: DEFINITION

In this contract, the following terms shall, unless otherwise specifically defined, have the following meanings:

 A. "Ore" means Iron Ore of Chinese Origin.

 B. "USD-U.S. Currency" means the currency of the United States of America.

 C. "MT i.e. Metric Ton" means a ton equivalent to 1,000 kilograms.

 D. "Wet Basis" means ore in its natural wet state. WMT means Wet Metric Ton.

 E. "Dry Basis" means ore dried at 105 degrees centigrade. DMT means Dry Metric Ton.

 F. "Platts Price" means the last announced price for iron ore at U-metal website.

 G. "SGS" means Société Générale de Surveillance S.A. Switzerland.

 H. "Surveyor" means any Internationally Recognized Inspection Agency.

 I. "CIQ" means Entry-Exit Inspection & Quarantine of the People's Republic of China (PRC).

 J. "Loadport" means the port of loading.

K. "Work days" are based on the calendar of the PRC.

L. "CFR" means Cost and Freight (Incoterms 2010).

CLAUSE 2: COMMODITY

Iron Ore Fines	Fe 60.00%
Country of Origin	the PRC
Port of Loading	Yangluo Port, Wuhan, the PRC
Port of Destination	_____

CLAUSE 3: DELIVERY QUANTITY AND DELIVERY PERIOD

The total delivery quantity shall be 400,000 WMT (+/- 10% at the Seller's option).

Delivery term: 26 months

Delivery method: 55 thousand tons every 110 days

CLAUSE 4: GUARANTEED SPECIFICATIONS

(A) Chemical composition (on dry basis)

Fe	61.00% Min.
Al_2O_3	1% Max.
SiO_2	5.53% Max.
Sulphur	0.11% Max.
Phosphorous	0.01% Max.
Free moisture loss at 105 degrees centigrade	10% max.

(B) Physical composition (on natural basis)

Size (0-10 mm)	Above 10 mm	2% Max.
	0-10 mm	97% Min.
	Total	100%

CLAUSE 5: PRICE OF IRON ORE FINES (Fe 61.00 Min)

The last stated Platts price for 60% at the time of delivery of 10,000 tons each month.

CLAUSE 6: PRICE ADJUSTMENT

BONUS AND PENALTY

The price of Iron Ore stipulated on Clause 5 shall be adjusted by the following bonuses and penalties:

1. IRON ORE FINES (Fe)

BONUS:

For each 1.00% of Fe above 60.00%, the base price shall be increased by USD 1.50 per DMT, Fraction Pro Rata.

PENALTY:

For each 1.00% of Fe below 60.00%, the base price shall be decreased by USD 1.00 per DMT,

Fraction Pro Rata.

The Buyer has the right to reject the cargo for Fe content below 56%.

2. IMPURITIES

If the composition of Ore in respect to Alumina (Al_2O_3), Silica (SiO_2), Sulphur (S) and Phosphorus (P) exceeds OR drops from the respective guaranteed maximum as set forth in Clause 4 herein above, the Buyer and the Seller shall accept delivery of Ore by imposing penalties and bonuses provided below, Fraction Pro Rata.

Al_2O_3 USD 0.05 per DMT for each 1.00% in excess of 1% guaranteed.

SiO_2 USD 0.05 per DMT for each 1.00% in excess of 5.53% guaranteed.

S (Sulphur)   USD 0.05 per DMT for each 0.01% in excess of 0.11% guaranteed.

P (Phosophorus) USD 0.05 per DMT for each 0.01% in excess of 0.08 guaranteed.

3. MOISTURE

In the event that the free moisture loss at 105 degrees centigrade exceeds the guarantee max. of 10% as set forth in Paragraph A of Clause 4 herein above the shipment, based on loadport survey, the Seller shall pay to the Buyer the freight for the corresponding quantity of the excess moisture over guaranteed maximum calculated on the basis of the final wet weight.

4. SIZE

In the event that the oversize quantity exceeds the respective guaranteed maximum as set forth in Paragraph B of Clause 4, the Seller shall pay the penalty at USD 0.25 per Dry Metric Ton, Fraction Pro Rata.

CLAUSE 7: PAYMENT

CASH

Advance payment: equivalent to 15% of the value of 400,000 tons of cargo, which must be paid by the Buyer in two stages, including the stage of the contract of 5% and the mine visit stage of 10%. (Pre-payment will be based on the latest Platts price at the time of the contract). The Buyer must pay the remaining balance in cash for each partial delivery of 55,000 tons of cargo on the basis of the latest Platts price at delivery, with penalties and rewards agreed in the contract with 14% of pre-payment. The shipment will be delivered to the Buyer with the presentation of the customs documents and inspection from the Seller's side. If the delivery is in the port's warehouse, the only evidence presented to the Buyer will be the SGS certificate.

CLAUSE 8: DOCUMENTATION

A. Certificate of Origin (One original and two copies) to be endorsed/issued by the PRC Chamber of Commerce and detailed loaded quantity, commodity, carrying vessel and suppliers name and address.

B. Certificate of Quality and Certificate of Weight of contracted goods in 3 (three) original and 3 (three) copies each issued by SGS/surveyor. The Certificate of Quality to show actual results of the test of chemical composition and all other tests as called for in this contract.

C. Invoice.

D. Bill of lading.

CLAUSE 9: WEIGHING

WEIGHING AT LOADPORT

At the loading port, the Seller at the Seller's expenses shall appoint SGS/surveyor to determine the weight of shipment of Iron Ore Fines by draft survey.

The weight of Ore as ascertained and certified by SGS/surveyor together with analysis shall be the basis of the Seller's Invoice. The Buyer or the Buyer's representative approved by the Seller shall have the right to be present at such weighing at the Buyer's expense at the loading port.

CLAUSE 10: SAMPLING AND ANALYSIS

At the time of loading of each shipment, SGS/surveyor shall take representative samples in accordance with the prevailing standards. The Buyer or the Buyer's representative approved by the Seller shall have the right to be present at such sampling at the Buyer's expense.

SGS/surveyor shall analyze Iron (Fe) content and other chemical composition on dry basis and physical composition on standard ISO 4701, and shall issue a certificate of such analysis.

The cost of such sampling and analysis shall be to the Seller's account. The analysis thus determined shall be the basis for the Seller's Invoice.

CLAUSE 11: ADVICE OF SHIPMENT

The Seller shall, upon completion of loading, advise the Buyer by phone/email/cable/fax the following:

    1. Contract No.;

    2. Name of Commodity;

    3. Invoice Value;

    4. Gross Weight & Loading Data;

    5. Bill of Lading No. & Date.

CLAUSE 12: FORCE MAJEURE

(A) In the event of delivery of all or any part of Ore under this contract being obstructed and/or delayed due to or resulting from cause or causes beyond the control of the Seller and the Buyer, such as war, hostility, military operation, civil commotion, sabotage, quarantine restriction, acts of Government, fire, floods, explosion, epidemics, blockades, revolutions, insurrection, mobilization, strikes, lockouts, riots, act of God, the Seller or the Buyer shall be relieved of the responsibility for performance of this contract as per Paragraph C hereinafter to the extent to which such performance has been obstructed.

(B) In the event that Force Majeure condition occurs prescribed in Paragraph A hereinabove, the party shall advise by cable the other party as soon as possible and then shall, within 2 (two) weeks after occurrence of such event, furnish the other party in writing with the particulars of the relevant event and documents explaining that its performance is prevented or delayed due to cause or causes as set forth in Paragraph A hereinabove and further shall furnish at the same time or at least within 2 (two) weeks after occurrence of such Force Majeure condition.

The party declaring a Force Majeure shall during the duration of such Force Majeure condition uses its best effort to resume the performance of its obligations under this contract with the least possible delay and such party shall always advise the other party of detailed progress of the event of Force Majeure and the prospect of settlement of such event and of the resumption of the performance of its obligations under this contract.

(C) In the event that the Seller or the Buyer is relieved of the responsibility for performance of this contract to the extent to which such performance has been obstructed and if approved by the other party, the time of delivery may be postponed for the duration of cause or causes hereinabove mentioned. If the duration of the postponement of this contract mentioned herein exceeds 3 (three) months, the party not declaring Force Majeure shall have the option to cancel this contract in respect of the undelivered quantity or extend the period of delivery by mutual agreement.

CLAUSE 13: PENALTY

In the event that the Buyer delays the time of the deposit during the time of the transaction in accordance with Clause 7, any delay in the provision of the product and the costs imposed on the Seller due to this delay will be borne by the Buyer.

CLAUSE 14: ARBITRATION

All disputes or differences whatsoever arising between the parties out of or relating to the construction, meaning and operation or effect of this contract or breach thereof shall be settled by Arbitration in accordance with the PRC Rules of Arbitration and the award made in pursuance thereof shall be binding on both parties. The venue of arbitration proceedings will be the PRC. The contract shall be governed and construed by the PRC Law.

CLAUSE 15: TITLE AND RISK

According to the Incoterms 2010, the CFR clause, the Seller pays for the carriage of the goods up to the named port of destination. Risk transfers to the Buyer when the goods have been loaded on board the ship in the country of export. If the delivery of the port floor is agreed, then the transfer of the SGS will end with the Seller's responsibility.

CLAUSE 16: LOSS OF CARGO

In the event of partial loss of cargo, the Bills of Lading weight and the analysis carried out by the SGS company on the cargo on loading port, shall be treated as final and shall form the basis of invoicing and payment. And the Seller shall not be liable for the same. In such situation, the Buyer shall claim the losses from the insurance company directly.

CLAUSE 17: VALIDATION AND AMENDMENT OF THE CONTRACT

This contract shall become effective when the duly authorized representatives of the Seller and the Buyer sign thereon. Any change, modification in or addition to the terms and conditions of this contract shall become effective when confirmed by both the Seller and the Buyer in writing.

CLAUSE 18: ENTIRETY OF CONTRACT

All proposals and negotiations if any, made prior to the date hereof are merged herein and no modifications or assignment shall be effective unless agreed upon by both the Seller and the Buyer.

CLAUSE 19: PROCEDURE

A. The Buyer and the Seller agree to the terms and sign/seal contract.

B. The Buyer pays a set percentage of the advance payment.

C. Shipment is effected as per contract.

CLAUSE 20: COMMISSION

A. % the Buyer Side...

B. % the Seller Side: It is announced at the stage of signing the contract.

CLAUSE 21: NOTICES

All communications referred to in this contract shall be in writing. And a copy of any document to the Buyer and the Seller remain a copy.

IN WITNESS WHEREOF

The Buyer and the Seller hereto have executed this Agreement in duplicate on the date first above written, each of them to be retained by the respective party of this Contract.

SELLER   BUYER

虽然这份合同的事由为铁矿细粉买卖，但涉及的冶金知识并不多，包含的冶金术语也不多。理解和翻译这份合同更多需要的是贸易领域的知识，以及对贸易英语文体特点的熟悉。这些特点包括但不限于：

为了文体正式使用古语词（如 whereof、hereto、herein、hereof 等）；

大量使用贸易术语（如 invoice、bill of lading、advance payment、certificate of origin、certificate of quality、delivery quantity、port of loading、port of destination 等）；

频繁使用 shall 表示义务（如 The cost of such sampling and analysis shall be to the Seller's account）；

为了严谨，并列使用意义相同或类似的词语（如 terms and conditions、ascertained and certified、obstructed and/or delayed、prevented or delayed、disputes or differences 等）；

频繁使用条件句，特别是 in the event that 引导的条件句（如 In the event that Force Majeure condition occurs prescribed in Paragraph A herein above）；

大量使用长句，特别是不可抗力条款，句子之长，为其他文体所少见。

冶金新闻英语主要报道冶金行业发生的事件或冶金市场的运行情况，涉及的冶金知识也不多，主要表现出新闻英语的特点。比如：

ArcelorMittal Acquiring 3 Million Metric Tons of Slabmaking Capacity in Brazil

07/28/2022—ArcelorMittal is acquiring Brazilian slab producer Companhia Siderúrgica do Pecém（CSP）for approximately US＄2.2 billion, the steelmaker said on Thursday.

In a statement, ArcelorMittal said CSP will add 3 million metric tons of slab capacity, expanding its footprint in the Brazilian market and opening the potential to ship slab to other ArcelorMittal plants. It also will allow the company to capitalize on a planned third-party investment to form a clean electricity and green hydrogen hub in Pecém.

"In CSP, we are acquiring a modern, efficient, established and profitable business which further enhances our position in Brazil and adds immediate value to ArcelorMittal. There is significant potential to decarbonize the asset given the state of Ceará's ambition to develop a low-cost green hydrogen hub and the huge potential the region holds for solar and wind power generation," said ArcelorMittal chief executive Aditya Mittal.

"CSP produces high-quality slabs and is cost competitive, ensuring its products are competitive domestically and for export. In the short term, we will continue to supply CSP's existing customer base in North and South America. However, what makes this acquisition so exciting is the medium-to-long-term potential and options it presents. As we continue to develop the downstream capabilities of our NAFTA and Brazilian businesses over the medium term, we have the option for CSP to become an important intra-group slab supplier. Over the longer term there is also the option to significantly increase its slab capacity and add rolling and finishing capabilities on a low-carbon emissions basis," he added.

A joint venture among iron ore miner Vale, Dongkuk Steel and POSCO, CSP has a single blast furnace and conveyor access to the Port of Pecém, a large, deepwater port located 10 km from the plant.

这篇新闻报道的是印度安赛乐米塔尔集团耗资近22亿美元收购了巴西一家板坯生产商，带有鲜明的新闻英语的特点，如：

采用倒金字塔结构（inverted-pyramid structure），信息的重要程度随语篇展开而递减；

大量使用直接引语以显客观；

为了简洁，在名词前使用较长修饰语（如 Brazilian slab producer Companhia Siderúrgica do Pecém、a planned third-party investment、a low-cost green hydrogen hub 等）；

使用一些短小有力的词语（如 add、open、ship、form、hold 等）；

大量使用现在时，等等。

与冶金贸易英语、冶金新闻英语不同，冶金科技英语涉及较多冶金科技知识，文体上表现出科技英语的特点。大致说来，根据读者对象的不同，冶金科技英语可分为两类：一类带有科普性质，面向的是普通大众，对读者的冶金科技素养要求不高，即使是对冶金科技了解不多的读者也可以理解；另一类由有冶金专业背景的人士甚至专家撰写，预期读者也是冶金专业人士，典型语篇如冶金科技论文或专著。请看下面两个例子：

【例1】

Steel is a metal alloy whose major component is iron, with carbon content between 0.02 and 1.7 percent by weight. Carbon is the most cost-effective alloying material for iron, but many other alloying elements are also used. Carbon and other elements act as a hardening agent, preventing dislocations in the iron atom crystal lattice from sliding past one another. Varying the amount of alloying elements and their distribution in the steel controls qualities such as the hardness, elasticity, ductility, and tensile strength of the resulting steel. Steel with increased carbon content can be made harder and stronger than iron, but is also more brittle.

The maximum solubility of carbon in iron is 1.7 percent by weight, occurring at 1,130°; higher concentrations of carbon or lower temperatures will produce cementite, which will reduce the material's strength. Alloys with higher carbon content than this are known as cast iron because of their lower melting point. Steel is also to be distinguished from wrought iron with little or no carbon, usually less than 0.035 percent. It is common today to talk about "the iron and steel industry" as if it were a single thing—it is today, but historically they were separate products.

Currently there are several classes of steels in which carbon is replaced with other alloying materials, and carbon, if present, is undesired. A more recent definition is that steels are iron-based alloys that can be plastically formed (pounded, rolled, and so forth).①

【例2】

In 2015, the global steel production obtained from blast furnace (BF) and electric arc furnace (EAF) cycles amounted to 1.11 billion tons. The demand for large quantities of high quality steel encouraged the creation of integrated steel mills based on the coupling of blast furnaces and basic oxygen furnaces (BOF). However, in recent years, environmental requirements have become more stringent and the electric arc furnace turns out to be a good alternative to integrated plants. On the other hand, EAF cycles will never completely replace integrated steel.

The increase of the steel scrap and hot briquetted iron (HBI) used in the BF-BOF route reduces the greenhouse gases (GHG) emissions involved in this technological route and is applicable to the existing BF-BOF plants in a reasonable timescale. Moreover, the increased scrap used in BOF does not require major changes in the steelmaking practices.

Furthermore, the increased availability and reduced cost of natural gas has prompted steelmakers to investigate the implementation of direct reduced iron (DRI) in their existing steelmaking operations to decrease operating costs and environmental impact. The addition of DRI in the blast furnace to increase the hot metal output and decrease the coke rate is well known. Unfortunately, this option has economic merits only if the downstream equipment of the steel plant is able to process the additional iron units into cast or rolled steel.②

这两个例子分别出自 *New World Encyclopedia* 和 *Journal of Iron and Steel Research International*。比较起来，前者在知识的深度上不及后者，专业术语的密度也不如后者，受过中等教育的人基本能理解前一个语篇，但要理解后一个语篇，就需要一定的专业背景，或临时补充足够的冶金流程相关知识。

---

① 摘自 *New World Encyclopedia*[EB/OL]. https://www.newworldencyclopedia.org/entry/Steel。

② Cosmo Di Cecca, et al. Thermal and chemical analysis of massive use of hot briquetted iron inside basic oxygen furnace[J]. *Journal of Iron and Steel Research International*, 2017 (24).

虽然有这些不同，两个语篇在文体特点上却没有多少区别，呈现的都是典型的科技英语的文体特点，包括：

大量使用科学术语，专业性突出，如前例中有 metal alloy、carbon content、alloying elements、hardening agent、dislocations、crystal lattice 等，后例中有 BF、BOF、EAF、integrated steel mill 等；

大量使用一般现在时，表示陈述的内容是客观真理，如前例中的 Steel is a metal alloy whose major component is iron, with carbon content between 0.02 and 1.7 percent by weight，后例中的 The increase of the steel scrap and hot briquetted iron (HBI) used in the BF-BOF route reduces the greenhouse gases (GHG) emissions involved in this technological route and is applicable to the existing BF-BOF plants in a reasonable timescale；

语言正式、规范，不使用口语体词语，更不使用俚语、俗语等，句子完整，很少有句子成分省略的情况；

表达质朴，不使用带有情感色彩的词语，非必要不使用形容词和副词，即使使用也以描述性形容词、副词为主，很少使用评价性形容词和副词[①]；

较多使用非谓语动词结构，以便表达简洁，如前例中有 preventing dislocations in the iron atom crystal lattice from sliding past one another，后例中有 obtained from blast furnace (BF) and electric arc furnace (EAF) cycles、to investigate the implementation of direct reduced iron (DRI)、used in the BF-BOF route、involved in this technological route 等。

考虑到冶金贸易英语、冶金新闻英语与冶金学科关系不大，翻译这些文本时不需要多少冶金知识，本教程将主要讨论冶金科技英语的翻译。

三、冶金科技英语翻译

冶金科技英语翻译指冶金科技语篇在英语、汉语之间的转换。要做好冶金科技英语翻译，不仅需要具备较高的英语、汉语理解和表达的能力，同时需要掌握冶金学科的知识。

一般来说，从事冶金科技英语翻译的人士要么是英语或翻译专业出身，要么是冶金专业出身。对于前者，难的是对冶金原料、产品、技术、流程等的了解；对于后者，难的是要具备较高的英语、汉语理解和表达能力。理想的冶金科技翻译人员应该既是冶金专家也

---

① 描述性、评价性是从语义上对形容词、副词进行的一种分类。描述性指事物自有的、客观的属性；评价性指带有主观的感觉，如 a red rose 中 red 是 rose 的自有客观属性，而 a lovely rose 中的 lovely 则含有说话人的评价。

是语言专家,但做到这点并不容易。

对于英语或翻译专业学生来说,冶金术语常是冶金科技翻译的第一个障碍。例如:

During pelletizing, the mixtures made from ultrafine (minus 0.074 mm or minus 200 mesh) iron-ore concentrates and binders of grain sizes far less than 1 mm are balled to form green pellets slightly larger than 6 mm but smaller than 15 mm in diameter.

此句中,像 iron-ore concentrates、binders、green pellets 等都可能成为翻译障碍。请看译文:

生产球团时,用极细的铁精矿粉(小于0.074毫米或小于200目)以及粒度远小于1毫米的黏结剂组成的混合物制成直径略大于6毫米但小于15毫米的生球。

再看一个例子:

A recent development is the bell-less top in which the charge is distributed from sealed bunkers at the furnace top by means of rotating chute.

不了解高炉生产的人看到这样一句话肯定会发蒙:bell-less top、the charge、sealed bunkers、rotating chute 都是些什么呢?即使看到这句的译文"高炉的最新发展是无料钟炉顶,借助于旋转溜槽把炉料从顶部密封料仓布入高炉内",也不一定清楚"无料钟炉顶""旋转溜槽"等是怎么回事。

词典确实能给我们帮助,但不一定能解决问题,这是因为再大的词典也有可能没有收录我们要查的词语。比如:

The kinds of coal, in increasing order of alteration, are lignite (brown coal—immature), sub-bituminous, bituminous, and anthracite (mature).

句中,对应 immature、mature 的汉语词语分别是"不成熟的"和"成熟的",但这里是否可以将 immature coal、mature coal 译成"不成熟的煤"("生煤")和"成熟煤"("熟煤")呢?笔者查遍了手头的各类词典,也没有准确答案。根据互联网上的资源,"生煤"指"没有烧透的煤"或"烟煤","熟煤"指炼焦时加入直立炭化炉中未经洗选的煤经过干馏后得到的固态产物,又称"不洗焦"。而从上下文来看,这里的 immature coal、mature coal 分别指

"煤化程度低的煤""煤化程度高的煤"①，因此最后将句子译成：

按改变程度由小到大煤炭可分为褐煤(褐色的煤——低煤化)、次烟煤、烟煤、无烟煤(高煤化)。

有时词典收录了我们要查的词语，但却给出了不止一种译法，比如：

Coal starts off as peat.

《英汉冶金工业词典》给出了 peat 的两种译法：泥煤、泥炭。② 到底该选择哪种译法呢？其实，"泥煤"和"泥炭"是同样一个东西的两种说法，译成"泥煤"或"泥炭"都是可以的。

开始形成的煤炭是泥炭/泥煤。

还比如：

Calcium has a greater affinity than iron for sulphur. Iron has a greater affinity than calcium for oxygen.

根据《英汉技术词典》，affinity 有如下意思：类似，相似，亲缘，共鸣，吸引；亲和力(性，势，能)，化合(亲和)力；仿射(相似)性。③
从这些译法中选择一个最准确的无疑需要学科知识：

钙对硫的亲和力比铁大，铁对氧的亲和力比钙大。

---

① 根据宋春青、张振春的《地质学基础》(第二版)(北京：高等教育出版社，1982：269-271)：煤的形成大致分成三个阶段，开始是腐解阶段(泥炭化阶段)，植物堆积在水下被泥沙覆盖，逐渐与氧气隔绝，由厌氧细菌参与，促使有机质腐烂分解。植物遗体中氢、氧等成分逐渐减少，碳的相对含量逐渐增加，生成泥炭。接着是煤化阶段(褐煤阶段)。在这个阶段，泥炭被沉积物覆盖，处于完全封闭的环境，细菌作用逐渐停止，泥炭被压缩、脱水和胶结，碳的含量进一步增加，成为褐煤。最后是变质阶段(烟煤、无烟煤阶段)，褐煤埋藏在地下较深位置时，受到高温高压的作用，水分和挥发成分减少，碳的含量相对增加，密度、比重、光泽和硬度等增加，成为烟煤。烟煤进一步变质成为无烟煤。无烟煤进一步变质可以成为石墨。总之，煤化过程就是碳的成分逐渐增加，氢、氧、硫等逐渐减少的过程。
② 《词典》编辑组. 英汉冶金工业词典[M]. 北京：冶金工业出版社，1999：1091.
③ 清华大学《英汉技术词典》编写组. 英汉技术词典[M]. 北京：国防工业出版社，1985：44.

有时，词典上给出的译法都不合适，我们必须依据上下文意思想出一个合适的译法。如：

As the oven is heated, the coal is cooked, so most of the volatile matter such as oil and tar is removed.

根据《英汉技术词典》，cook 对应汉语中的"蒸煮，烧，熬（热）炼，烹调；杜撰，篡改，伪造，捏造，虚报；计划，设计；炊事员，厨师"①。
从语境可以看出，这里的 is cooked 就是 is heated 的意思，因此可将其译为"受热"：

给焦炉加热，煤炭受热，大部分挥发物质被除去，例如油和焦油。

又比如：

The majority of coke produced in the United States comes from wet-charge, by-product coke oven batteries.

查《英汉冶金工业词典》，wet charge 有"生料""湿料"两个译法②，但这两个词语用在译文中都不够清楚，因此根据语境，将其译为"湿煤装料"：

美国生产的大部分焦炭来自湿煤装料、副产炼焦炉组。

解决了学科知识和冶金术语的问题后，还有一个更重要的问题：什么样的译文才是好的冶金科技译文？或者说，判断冶金科技翻译质量应遵循什么标准呢？

(一) 冶金科技英语翻译标准

翻译标准"指翻译活动必须遵循的标准，是衡量译文质量的尺度，是翻译工作者不断努力以期达到的目标。切实可行的标准对发挥翻译功能、提高翻译质量具有重要的意义"③。
应遵循什么样的翻译标准是所有译者翻译之前必须回答的首要问题，因为只有明确了

---

① 清华大学《英汉技术词典》编写组. 英汉技术词典[M]. 北京：国防工业出版社，1985.
② 《词典》编辑组. 英汉冶金工业词典[M]. 北京：冶金工业出版社，1999：1623.
③ 方梦之. 中国译学大辞典[M]. 上海：上海外语教育出版社，2011：68.

翻译标准，译者的努力才有方向。关于翻译标准的讨论在中西方都有悠久的历史，并且出现了很多富有真知灼见的论断，其中比较著名的有如下这样一些：

严复 译事三难：信、达、雅。求其信，已大难矣！顾信矣，不达，虽译，犹不译也，则达尚焉。……《易》曰："修辞立其诚。"子曰："辞达而已矣！"又曰："言之无文，行之不远。"三者乃文章正轨，亦即为译事楷模。故信、达而外，求其尔雅。①

鲁迅 无论什么，我至今主张"宁信而不顺"。自然，这所谓"不顺"，决不是说"跪下"要译作"跪在膝之上"，"天河"要译作"牛奶路"的意思，乃是说，不妨让译文不像吃茶淘饭一样几口可以咽完，却必须费些力气来咀嚼。②

林语堂 翻译的标准问题大概包括三个方面。……第一是忠实标准，第二是通顺标准，第三是美的标准。③

茅盾 对于一般翻译的最低限度的要求，至少应该是用明白畅达的译文，忠实地传达原作的内容。但对于文学翻译，仅仅这样要求还是不够的。……文学的翻译是用另一种语言，把原作的艺术意境传达出来，使读者在读译文的时候能够像读原作时一样得到启发、感动和美的感受。④

傅雷 以效果而论，翻译应当像临画一样，所求的不在形似而在神似。⑤

钱锺书 文学翻译的最高理想可以说是"化"。把作品从一国文字转变成另一国文字，既能不因语文习惯的差异而露出生硬牵强的痕迹，又能完全保存原作的风味，那就算得入于"化境"。⑥

亚历山大·弗雷泽·泰特勒（Alexander Fraser Tytler） Ⅰ. The Translation should give a complete transcript of the ideas of the original work. Ⅱ. The style and manner of writing should be of the same character with that of the original. Ⅲ. The Translation should have all the ease of original composition.⑦

约翰·坎尼森·卡特福德（John C. Catford） Translation may be defined as the replacement of textual material in one language (SL) by equivalent textual material in another language (TL).⑧

---

① 严复. 天演论·译例言[M]. 北京：商务印书馆，1902.
② 鲁迅. 关于翻译的通信[M]//鲁迅全集：第4卷. 北京：人民文学出版社，2005.
③ 林语堂. 论翻译[M]//林语堂全集. 北京：中国社会科学出版社，1984.
④ 茅盾. 为发展文学翻译事业和提高翻译质量而奋斗[R]. 1954-08-19.
⑤ 傅雷. 高老头·重译本序[M]//傅雷译文集. 北京：人民文学出版社，1981.
⑥ 钱锺书. 林纾的翻译[M]//钱锺书散文. 北京：人民文学出版社，1997.
⑦ Tytler, A. F. *Essay on the Principles of Translation*[M]. Edinburgh：J. Walter，1790.
⑧ Catford, J. C. *A Linguistic Theory of Translation：An Essay in Applied Linguistics*[M]. London：Oxford University Press，1965.

尤金·A.奈达(Eugene A. Nida)　Translating consists in reproducing in the receptor language the closest natural equivalent of the source language message, first in terms of meaning and secondly in terms of style.①

这些标准无疑对于翻译活动具有一定的指导价值，但是翻译活动多种多样，翻译目的各不相同，试图用一套标准来规范所有翻译行为注定会失败。且不说"口译"和"笔译"差别何其之大，即使同为笔译，文学翻译和科技翻译的差异也是非常明显的。文学翻译强调文学形象、审美效果的传达，科技翻译注重信息传递的完整和准确，显然不应遵循同样的翻译标准。

剔除以自娱自乐为目的的翻译——这类翻译可以随意发挥。翻译事实上是一种产品，译者是生产商，出版社是批发商，书店（线下、线上）是零售商，委托人、读者等是顾客。顾客需求多种多样，翻译类型也必须多种多样，规范翻译的标准也必须不同，正如夹克、牛仔裤、羽绒服虽然都归属服装，但生产时却遵循不同的标准一样。

所以，首先，翻译活动需要的不是一个单一标准，而是一个由很多标准组成的标准体系，以规范不同类型翻译产品的生产。

其次，既然是标准，就不能太理想化，不能将绝大多数译文排斥在外，像"雅""化境""神似"之类的标准，就不够实用。这与制定服装标准的道理是一样的：如果一种服装的质量标准无法适用于市场上大部分服装，那么问题很可能不在服装，而在服装标准。

再次，既然是标准，就应该具有可执行性，即译者、读者等可以对照标准轻松判断译文是否满足要求。然而，上面这些标准大多过于笼统，可执行性不足。例如"忠实"，什么样的译文才算"忠实"呢？又比如"equivalent textual material"，什么样的译文才是"equivalent"呢？理想的标准需要像实物产品标准一样，提供更详细的描述。

翻译既然是产品，满足顾客即读者的需求应该是翻译的第一要义，而读者的需求通常又与文本的功能相关。

语言学研究发现，每一种文本都有其功能，这些功能主要包括：信息功能(informative)、交际功能(interpersonal)、娱乐功能(recreational)、寒暄功能(phatic)、情感功能(emotive)、施为功能(performative)、元语言功能(metalingual)等。虽然一种文本可以有多重功能，但通常以一种功能为主，如文学作品以娱乐读者为主，科技文本以传递信息为主。

除少数例外（如将私人信件当作文学作品来翻译，私人信件本来是用于交际和表达情感，而当作文学作品来翻译时，译文便具有了文学作品的功能），翻译的目的大多是再现文本的本来功能。同其他类型的科技文本一样，冶金科技文本的主要功能首先是传递冶金

---

① Nida, E. A. *The Theory and Practice of Translation*[M]. Leiden: E. J. Brill, 1964.

科技信息，因此确保信息的准确与完整是冶金科技翻译的首要任务，也是判断冶金科技翻译质量的首要标准。尤其是冶金科技文本中常含有数字，翻译时千万不能弄错，否则可能会造成重大的经济损失。

其次，作为科技文本的一种，冶金科技文本的重要特点之一便是其科技性，表现在大量使用科技术语上，这种科技性在翻译中必须得到再现，这是对冶金科技翻译的合理要求。例如：

The best **grades** of ore contain over 60% iron. Lesser grades are treated, or refined, to remove various contaminants before the ore is shipped to the blast furnace.

译文一：最好**等级**的矿石含铁可达60%以上。较低**等级**的矿石在送进高炉前需要处理或加工，去除各种杂质。

译文二：最高**品位**的矿石含铁可达60%以上。较低**品位**的矿石在送进高炉前需要处理或加工，去除各种杂质。

Grade虽然有"等级"的含义，但用在描述铁矿石上，正确的说法是"品位"。

还比如：

Before carbonization, the selected coals from specific mines are **blended**, pulverized, and oiled for proper bulk density control.

译文一：碳化前，从特定煤矿选来的煤炭需要**混合**、粉碎和加油，以达到合适的堆密度。

译文二：碳化前，从特定煤矿精选煤炭进行**配煤**、粉碎和加油，以达到合适的堆密度。

Blend有"混合"的意思，但用在混合煤炭以便取各煤种之长炼焦上，正确的说法是"配煤"。

为了保留译文的科技性，翻译时千万不能将科技术语当作通用词汇。如：

Sponge iron is chemically much more active than steel or iron in the form of millings, **borings**, **turnings** or wire.

这里的boring和turning不是常见的"无聊的"和"旋转、转向"的意思，而分别是"钻粉""削屑"的意思。因此句子应该译作：

当海绵铁以**磨粉**、**钻粉**、**削屑**或金属丝形式存在时，海绵铁比钢和铁化学性质更活泼。

再次，冶金科技文本要么用于学术交流，要么用于冶金科技知识普及，目的都很严肃，自然地要求译文语言也要正式，这算冶金科技翻译的第三条标准。例如：

Called "puddling", this was highly skilled work, but was also **hot, strenuous, and dangerous**.

译文一：这种工作被称为"搅练"，技术性很强，但又热又累又危险。

译文二：这种工作被称为"搅练"，技术性很强，但**工作环境酷热，劳动强度大，危险性高**。

与译文一相比，译文二语言正式，更符合冶金科技文体的语言使用特点。

最后，无论是学术交流还是知识普及，效率都是关键，因此冶金科技文本大多语言表达简洁。同时，既然是学术语言，其表达应当客观，应少用文学化、情绪化的语言。所以，语言的简洁和质朴是冶金科技翻译的第四条标准。例如：

Sponge iron may be produced as a granular material or as a sintered mass, depending upon the methods of manufacture.

译文一：取决于生产工艺，海绵铁可以以粒状材料形式也可以以烧结块形式被生产出来。

译文二：取决于生产工艺，海绵铁生产出来后可呈粒状，也可能是烧结起来的一块。

BOF heat sizes in the U. S. are typically around 250 tons, and tap-to-tap times are about 40 minutes, of which 50% is "blowing time".

译文一：在美国，典型的BOF每炉生产钢大约250吨，需要时间大约为40分钟，其中一半为吹氧时间。

译文二：在美国，典型BOF每炉产钢约250吨，需时约40分钟，其中一半为吹氧时间。

Two types, "soft" and "hard" burned lime, are available.

译文一：有两种类型的烧石灰："软的"和"硬的"。

译文二：有"软""硬"两种类型的烧石灰。

上述例句中，译文二都比译文一简洁。

Ladle additives are available to reduce the iron oxide level in the slag, but nothing can be done to alter the phosphorus.
译文一：钢包添加剂可以降低渣中的铁氧化物水平，但对于磷则无可奈何/束手无策。
译文二：钢包添加剂可以降低渣中的铁氧化物水平，但无法改变磷的含量。

虽然 nothing can be done 可以被译为"无可奈何"或"束手无策"，但两个成语都不够质朴，不如简单地译为"无法"。

当然，所有的文章都应该通顺，译文的通顺是应有之义，所以这里就不再将通顺视作冶金科技翻译的标准之一了。

（二）冶金科技英语翻译方法

虽然冶金科技英语文本在内容和表达上有别于其他文本，但冶金科技英语翻译所采用的常用方法与其他类型文本的翻译并无二致，无非音译、直译、意译、变译等①。

1. 音译

音译主要用于专有名词或源语中有而目的语中不存在的术语的翻译，如：Andrade creep 安德雷德蠕变、Bessemer converter 贝塞麦转炉、Colonial 科洛尼亚尔（耐蚀铬镍合金钢）。

2. 直译

直译②主要用于源语和目的语在表达方式、句法结构上差别不大的情形，译者按原文直接翻译即可，无须采取任何变换措施。直译主要用于简单句，如：

A blast furnace charge consists of coke, iron ore or sinter, and limestone.
一炉高炉炉料包括焦炭、铁矿石或烧结料以及石灰石。

---

① 严格说来，音译、直译、意译、变译不能相提并论，因为音译发生在词汇层面，直译和意译一般发生在句子层面，而变译一般发生在段落或篇章层面，但习惯上都将它们并列讨论，均视之为翻译方法。
② 直译、意译概念虽由来已久，但对于这两个概念学界并没有一致认可的定义。一般认为，尽管直译、意译也可用于词汇（主要是复合词语、词组，如 Achilles' heel 译成"阿喀琉斯之踵"或"阿喀琉斯脚后跟"是直译，而译成"致命弱点"则是意译，单个词语则不存在直译、意译）的翻译，但通常指句子层面是否发生了改变。

Non-ferrous metals don't contain iron, for example aluminum, copper and titanium.

有色金属不含铁，如铝、铜和钛。

3. 意译

由于中、英两种语言及其背后的文化差异太大，中英互译时大多不能直译，必须进行词语或结构的调整，也就是意译。这些调整主要包括：增词（Amplification）、减词（Omission）、词性/语态转换（Conversion）、词序调整（Inversion）、正说反译/反说正译（Negation）、长句拆译（Division）、分清主从（Subordination）等。请看下面的例句：

This ore is either Hematite（Fe_2O_3）or Magnetite（Fe_3O_4）and the iron content ranges from 50% to 70%. This iron rich ore can be charged directly into a blast furnace without any further processing. Iron ore that contains a lower iron content must be **processed or beneficiated** to increase its iron content.

矿石或者是褐铁矿或者是磁铁矿，含铁量在50%—70%之间。这些含铁量高的矿石可以不经处理直接装入高炉，而含铁量较低的矿石则必须经过**选矿**等处理增加其含铁量。

相对于原文，译文增加了"而"和"则"，这是典型的增词，而原文中的 processed or beneficiated 则按意思被调整成了"选矿等处理"。

Unlike the open hearth, the BOF operation is conducted almost "in the dark" using mimics and screens to determine vessel inclination, additions, lance height, oxygen flow, etc.

跟平炉不同的是，BOF操作基本是"在黑暗中"进行的，使用模拟设备和显示屏来控制炉体的倾斜、**炉料的**添加、氧枪的高度和吹氧等。

为了补足意思，译文增加了"炉料的"三个字，也是增词不增意。

The raw ore is removed from the earth and sized into **pieces** that range from 0.5 to 1.5 inches.

原矿从地球开采出来，被加工成0.5—1.5英寸大小。

原文中虽然有 pieces，但不译并不影响原文意思的传达，减词有其必要。

第一课 概　　论

While **the process** of producing low carbon iron directly from ore is theoretically attractive and appears more logical than indirect processes, direct processes have so far failed in competition with indirect methods.

尽管直接从矿石生产低碳钢理论上吸引人，听起来也比间接法符合逻辑，但迄今直接法在与间接法的竞争中一直处于下风。

原文中的 process 没有译出，但意思不减。

The sulfur containing compounds report to the slag; however, unless the sulfur-rich slag is skimmed before the hot metal is poured into the BOF, the sulfur actually charged will be well above **the level** expected from the metal analysis.

含硫化合物进入炉渣。然而，除非在铁水倒进 BOF 之前将富含硫的炉渣撇去，否则实际进入转炉的硫还是会远远高于预期。

原文中 the level 同样没有译出，但不影响意思表达。

Once a blast furnace is started, it will continuously run for four to ten years with only short stops to **perform planned maintenance**.

高炉一旦开始生产，就会连续运转四到十年，只是间隔短时停产，以便**按计划开展检修**。

原文中 perform planned maintenance 本应译为"开展计划检修"，但读来别扭，采用词性转换，改译成"按计划开展检修"就顺畅得多。

Pellets are produced from this lower iron content ore. **This ore is crushed and ground into a powder** so that the waste material called gangue can be removed.

球团矿就是用这种含铁量低的矿石生产的。**将矿石破碎磨成粉**，以便去除被称为脉石的无用物质。

原文中动词使用了被动语态，翻译时按照汉语表达习惯将其转换成了主动语态，这属于语态转换。

The inertinite group includes fusinite, most of which is fossil charcoal, derived from ancient peat **fires**.

23

惰质组包括丝质体，大多数丝质体为木炭化石，产生于古代的泥煤火焚。

原文 fires 是名词，被译成了主谓词组"火焚"，这属于词性转换。

英、汉两种语言差异巨大，具体表现在语音、词语、句子、篇章等各个层面，其中句子结构层面的差异对英汉互译影响最大。学界普遍认为：汉语是意合语言，英语是形合语言。汉语句子属语义型，是依靠内在语义组织的；英语句子属句法型，是依据句法结构组织的。大部分英语句子必须有主谓结构，主谓结构是句子的主干。通过各种关联词和语法关系，以及 it 和 there 的广泛运用，各种从句、短语被附加在主干上，形成有条有理、共干多枝的树形结构。而汉语句子的主谓成分和其他语法成分的区分不明显，担任主谓成分的词、词组或结构也比较灵活。主语可以是名词性词组、形容词性词组，甚至动词性词组，谓语也是如此。另外，汉语的主谓成分没有形态变化，主谓之间不存在一致性关系，主语和谓语的搭配也不一定符合英语的表达逻辑，甚至有许多汉语句子没有主语或谓语，而是按时间或逻辑顺序，层层铺开，呈线形推进，形似竹竿。

英、汉两种语言在句子结构层面的巨大差异意味着，英、汉互译时常常需要进行词序调整（Inversion）。例如：

The precise filling order is developed by the blast furnace operators to carefully control gas flow and chemical reactions inside the furnace.

高炉操作工根据严格控制炉气流动和高炉内化学反应的要求，确定具体的布料顺序。

原句主语是 The precise filling order，这是英语中常见的物称主语，后面跟着的是被动语态。但汉语中常用有生命的人或动物作主语，少用主动语态，因此翻译时作了词序的调整。另外，汉语中常将目的放在前面，行为放在后面，这与英语相反，因此，to carefully control gas flow and chemical reactions inside the furnace 被译在了动作 develop the precise filling order 的前面。

In metalworking, rolling is a metal forming process in which metal stock is passed through one or more pairs of rolls to reduce the thickness and to make the thickness uniform.

金属加工中，金属原料通过一对或多对轧辊以减小厚度并使厚度均匀的金属成型工艺被称为轧制。

英语常用后置修饰语，汉语喜欢用前置修饰语，所以翻译时作了词序调整。

正说反译/反说正译(Negation)指用目的语中的否定句来译源语中的肯定句,而用目的语中的肯定句来译源语中的否定句。使用这种翻译方法,是因为英汉两种语言的肯定形式和否定形式以及表达肯定意义和否定意义的方式存在差异。汉语中常用的否定词如"不、无、非、没、未、勿、毋、莫、否、别"等,构词能力不强,常以结构否定的形式加在其他词汇前面表达否定。相比之下,英语的否定形式要复杂得多,有完全否定(如 no、not、none、never、nothing)、绝对否定(如 not at all、by no means、in no way 等)、半否定(如 hardly、scarcely、seldom 等)、含否定意义的词语(如 fail、defy、deny、lack、refrain)以及由含否定意义的前后缀(如 ab-、in-、im-、-less)构成的词语等。① 例如:

It is important to cast the furnace at the same rate that raw materials are charged and iron/slag produced so liquid levels can be maintained in the hearth and below the tuyeres.
出铁速度必须与原材料装料速度及渣/铁生成速度相同,以便液面在炉缸中保持不变,低于风口。

Those that were developed to meet purely local conditions with some degree of success might prove to be completely impractical under other conditions involving, for example, different fuels and different raw materials.
纯粹为了满足当地生产条件而开发的工艺在当地获得了某种成功,但在不同的条件下,比如不同的燃料和原材料,可能就会完全不实用。

前面说过,英语是形合的语言,英语句子中几乎每一个成分都可以加上修饰语,这使得一些英语句子呈现极其复杂的结构。然而,汉语句子大多以意驱动,结构简单,句子与句子之间按时间或逻辑顺序排列,呈小句接小句的流水状结构。因此英译汉时,经常需要将复杂的英语句子拆分成较短的小句,这便是长句拆译(Division)。例如:

The sulfur containing compounds report to the slag; however, unless the sulfur-rich slag is skimmed before the hot metal is poured into the BOF, the sulfur actually charged will be well above the level expected from the metal analysis.
含硫化合物进入炉渣。然而,除非在铁水倒进 BOF 之前将富含硫的炉渣撇去,否则实际进入转炉的硫还是会远远高于预期。

The first electric arc furnaces were developed by Paul Héroult, of France, with a

---

① 朱徽. 汉英翻译教程[M]. 重庆:重庆大学出版社,2004:87-88.

commercial plant established in the United States in 1907.

第一座电弧炉是法国人保罗·埃鲁研制的。1907 年，一座商用电弧炉冶炼厂在美国建成。

分清主从(Subordination)指将源语句子中的并列成分分别翻译成目的语中的主干成分和从属成分，以区分信息的轻重。分清主从主要用于汉译英，这是因为汉语没有主从之分，只有英语句子才分主干成分和从属成分。例如：

1856 年 8 月，一位名叫亨利·贝塞麦的英国人**公布了**他的炼钢方法，这个工艺最终能把钢的成本**降低到**原来的七分之一，更重要的是**使得**大量生产钢**成为了可能**。

In August 1856, an Englishman, Henry Bessemer, **made public** the description of his process of steelmaking which eventually **reduced** the price of steel **to** about a seventh of its former cost and more important still, **made it possible** to produce steel **in large quantities**.

原文中"公布了"和"降低到"所在的两个句子是并列关系，译成英语时，根据信息的重要程度，将"公布了" made republic 用作主句的谓语，而将"降低到"译成 which eventually reduced… 定语从句，主次轻重一目了然。

还比如：

平炉工艺**曾经几乎占**粗钢产量的 100%，**现今已缩减到**可忽略的程度。

The open-hearth process, **once responsible for** almost 100% of raw steel production, **has now dwindled to** negligible proportions.

原句中，"曾经几乎占"和"现今已缩减到"是并列关系，但考虑到信息的重要程度，将"现今已缩减到" once responsible for 当作了从属成分，而将"曾经几乎占" has now dwindled to 作为了句子的主干成分。

4. 变译

虽然变译实践非常久远，但作为一个学术概念，变译是国内学者黄忠廉最先提出的。黄忠廉认为，"变译是译者根据读者的特殊需求采用扩充、取舍、浓缩、阐释、补充、合并、改造等变通手段摄取原作中心内容或部分内容的翻译活动"。黄忠廉将变译大致分成了摘译、编译、译述、缩译、综述、述评、译评、改译、阐译、译写、参译等种类。[①]

很明显，所谓变译，就是翻译和写作的融合。普通的翻译注重形式和内容对原文的忠实，但有时候，读者需要的并不是形式和内容完全忠实于原文的译文，此时采用变译便有

---

① 黄忠廉. 翻译变体研究[M]. 北京：中国对外翻译出版公司，2000：5-6.

其必要了。

变译是一种翻译策略，而非具体的翻译方法。变译是在宏观层面对原文进行改造，而音译、直译、意译是在微观层面进行处理。变译可以直接基于原文展开，也可以先将原文按传统方式翻译成目的语，然后在译文基础上进行扩充、取舍、浓缩、阐释、补充、合并等。由于变译需要的技能主要涉及写作，这里就不举例展开了，感兴趣者可进一步阅读黄忠廉所著的《翻译变体研究》和《变译理论》或相关论文。

◎ 练习一　思考并回答下列问题。

1. 冶金科技英语具有哪些特点？如何才能学好冶金科技英语？
2. 冶金科技英语翻译应遵循什么原则？为什么要遵循这些原则？
3. 冶金科技英语翻译的方法有哪些？都用于什么情况？
4. 作为一项职业，冶金科技英语翻译在中国有无前途？为什么？
5. 如何才能成长为一名合格的冶金科技英语翻译？

第二课　铁

> When iron was found, the trees began to tremble, but the iron reassured them: "Let no handle made from you enter into anything made from me, and I shall be powerless to injure you."
>
> —*Genesis Rabbah* 5, *Tales and Maxims from the Midrash*[①]

一、科技术语

alloy A metal that is made by mixing two or more types of metal together.（合金）

ammonia A colorless liquid or gas with a strong, sharp smell. It is used in making household cleaning substances.（氨；氨水）

anthracite A type of very hard coal that burns slowly, producing a lot of heat and very little smoke.（无烟煤）

antiseptic A substance that helps to prevent infection in wounds by killing bacteria.（防腐剂；抗菌剂）

beneficiation The treatment of raw material (as iron ore) to improve physical or chemical properties, especially in preparation for smelting.（富集；选矿）

blast furnace A furnace in which combustion is intensified by a blast of air, especially a furnace for smelting iron by blowing air through a hot mixture of ore, coke, and flux.（高炉）

calcium carbonates A white crystalline salt occurring in limestone, chalk, marble, calcite, coral, and pearl: used in the production of lime and cement.（碳酸钙）

carbon dioxide A gas breathed out by people and animals from the lungs or produced by burning carbon.（二氧化碳）

carbon monoxide A colorless, odorless, very toxic gas (CO) that is formed as a product of the incomplete combustion of carbon or a carbon compound.（一氧化碳）

[①] Samuel Rapaport. *Tales and Maxims from the Midrash*. Noida: Amity EBooks, 2016.

chromium A hard, shiny, metallic element, used to make steel alloys and to coat other metals.(铬)

cinder A small piece of ash or partly burnt coal, wood, etc. that is no longer burning but may still be hot.(灰烬;炉渣)

coke Carbon fuel produced by distillation of coal.(焦炭)

cresol An aromatic compound derived from phenol, existing in three isomeric forms: found in coal tar and creosote and used in making synthetic resins and as an antiseptic and disinfectant; hydroxytoluene.(甲酚)

element A substance such as gold, oxygen, or carbon that consists of only one type of atom.(元素)

explosive A substance or device that can cause an explosion.(炸药;爆炸物)

heirloom An ornament or other object that has belonged to a family for a very long time and that has been handed down from one generation to another.(传家宝;祖传物)

herbicide A chemical that is poisonous to plants, used to kill plants that are growing where they are not wanted.(除莠剂;除草剂)

iron content Iron content tells the amount of iron in iron oxides.(铁含量)

iron oxide A compound of oxygen and iron.(铁氧化物)

limestone A type of white stone that contains calcium, used in building and in making cement.(石灰岩)

manganese A grayish white metal that is used in making steel.(锰)

melting point The melting point of a substance is the temperature at which it melts when you heat it.(熔点)

meteorite A large piece of rock or metal from space that has landed on Earth.(陨石)

mine When a mineral such as coal, diamonds, or gold is mined, it is obtained from the ground by digging deep holes and tunnels.(采掘)

molybdenum A very hard ductile silvery-white metallic element occurring principally in molybdenite: used mainly in alloys, esp. to harden and strengthen steels.(钼)

nickel A silver-colored metal that is used in making steel.(镍)

ore-enrichment The act of enriching iron content in iron ores.(矿石富集)

pelletize To form or shape (a substance) into pellets.(使成颗粒状;球团工艺)

pesticide Chemicals that farmers put on their crops to kill harmful insects.(杀虫剂)

pharmaceutical Connected with the industrial production of medicines.(制药的)

phenol A poisonous white chemical. When dissolved in water it is used as an antiseptic and disinfectant, usually called carbolic acid.(酚;苯酚;石碳酸)

pig iron Crude iron produced in a blast furnace and poured into moulds in preparation for making wrought iron, steels, alloys, etc. (生铁)

raw material Material on which a particular manufacturing process is carried out. (原材料)

refractory Difficult to fuse, corrode, or draw out; especially capable of enduring high temperature. (耐火的;耐热的)

sinter To form large particles, lumps, or masses from (metal powders or powdery ores) by heating or pressure or both. (烧结工艺)

slag The fused material formed during the smelting or refining of metals by combining the flux with gangue, impurities in the metal, etc. It usually consists of a mixture of silicates with calcium, phosphorus, sulphur, etc. (熔渣;炉渣)

smelt To extract a metal from an ore by heating. (熔炼)

solvent A liquid that can dissolve other substances. (溶剂)

toluene A colorless volatile flammable liquid with an odor resembling that of benzene, obtained from petroleum and coal tar and used as a solvent and in the manufacture of many organic chemicals. (苯)

tungsten A grayish-white metal. (钨)

tuyère Nozzle through which hot air is carried to the furnace. (鼓风口)

wrought iron A type of iron that is easily formed into shapes and is used especially for making gates, fences, and furniture. (熟铁)

二、英语原文

Iron

Background

1 Iron is one of the most common elements on Earth. Nearly every construction by humans contains at least a little iron. It is also one of the oldest metals and was first fashioned into useful and ornamental objects at least 3,500 years ago.

2 Pure iron is a soft, grayish-white metal. Although iron is a common element, pure iron is almost never found in nature. The only pure iron known to exist naturally comes from fallen meteorites. Most iron is found in minerals formed by the combination of iron with other elements. Iron oxides are the most common. Those minerals near the surface of the earth that

have the highest iron content are known as iron ores and are mined commercially.

3 Iron ore is converted into various types of iron through several processes. The most common process is the use of a blast furnace to produce pig iron which is about 92%–94% iron and 3%–5% carbon with smaller amounts of other elements. Pig iron has only limited uses, and most of this iron goes on to a steel mill where it is converted into various steel alloys by further reducing the carbon content and adding other elements such as manganese and nickel to give the steel specific properties.

History

4 Historians believe that the Egyptians were the first people to work with small amounts of iron, some five or six thousand years ago. The metal they used was apparently extracted from meteorites. Evidence of what is believed to be the first example of iron mining and smelting points to the ancient Hittite culture in what is now Turkey. Because iron was a far superior material for the manufacture of weapons and tools than any other known metal, its production was a closely guarded secret. However, the basic technique was simple, and the use of iron gradually spread. As useful as it was compared to other materials, iron had disadvantages. The quality of the tools made from it was highly variable, depending on the region from which the iron ore was taken and the method used to extract the iron. The chemical nature of the changes taking place during the extraction were not understood; in particular, the importance of carbon to the metal's hardness. Practices varied widely in different parts of the world. There is evidence, for example, that the Chinese were able to melt and cast iron implements very early, and that the Japanese produced amazing results with steel in small amounts, as evidenced by heirloom swords dating back centuries. Similar breakthroughs were made in the Middle East and India, but the processes never emerged into the rest of the world. For centuries the Europeans lacked methods for heating iron to the melting point at all. To produce iron, they slowly burned iron ore with wood in a clay-lined oven. The iron separated from the surrounding rock but never quite melted. Instead, it formed a crusty slag which was removed by hammering. This repeated heating and hammering process mixed oxygen with the iron oxide to produce iron, and removed the carbon from the metal. The result was nearly pure iron, easily shaped with hammers and tongs but too soft to take and keep a good edge. Because the metal was shaped, or wrought, by hammering, it came to be called wrought iron.

5 Tools and weapons brought back to Europe from the East were made of an iron that had been melted and cast into shape. Retaining more carbon, cast iron is harder than wrought iron and will hold a cutting edge. However, it is also more brittle than wrought iron. The European

iron workers knew the Easterners had better iron, but not the processes involved in fashioning stronger iron products. Entire nations launched efforts to discover the process.

6   The first known European breakthrough in the production of cast iron, which led quickly to the first practical steel, did not come until 1740. In that year, Benjamin Huntsman took out a patent for the melting of material for the production of steel springs to be used in clockmaking. Over the next 20 years or so, the procedure became more widely adopted. Huntsman used a blast furnace to melt wrought iron in a clay crucible. He then added carefully measured amounts of pure charcoal to the melted metal. The resulting alloy was both strong and flexible when cast into springs. Since Huntsman was originally interested only in making better clocks, his crucible steel led directly to the development of nautical chronometers, which, in turn, made global navigation possible by allowing mariners to precisely determine their east/west position. The fact that he had also invented modern metallurgy was a side-effect which he apparently failed to notice.

Raw Materials

7   The raw materials used to produce pig iron in a blast furnace are iron ore, coke, sinter, and limestone. Iron ores are mainly iron oxides and include magnetite, hematite, limonite, and many other rocks. The iron content of these ores ranges from 70% down to 20% or less. Coke is a substance made by heating coal until it becomes almost pure carbon. Sinter is made of lesser grade, finely divided iron ore which, is roasted with coke and lime to remove a large amount of the impurities in the ore. Limestone occurs naturally and is a source of calcium carbonate.

8   Other metals are sometimes mixed with iron in the production of various forms of steel, such as chromium, nickel, manganese, molybdenum, and tungsten.

The Ore Extraction and Refining Process

9   Before iron ore can be used in a blast furnace, it must be extracted from the ground and partially refined to remove most of the impurities.

10   Historically, iron was produced by the hot-blast method, or later, the anthracite furnace. Either way, the fundamental activity in iron making involved a worker stirring small batches of pig iron and cinder until the iron separated from the slag. Called "puddling", this was highly skilled work, but was also hot, strenuous, and dangerous. It required a lot of experience as well as a hearty constitution. Puddlers were proud, independent, and highly paid.

11   Puddlers founded the first trade union in the iron and steel industry, the Sons of Vulcan, in Pittsburgh in 1858. In 1876, this union merged with three other labor organizations

to form the Amalgamated Association of Iron and Steel Workers. This was the union that Andrew Carnegie defeated in the Homestead Strike of 1892, leaving the union in shambles and the industry essentially unorganized until the 1930s.

Extraction

12   Much of the world's iron ore is extracted through open pit mining in which the surface of the ground is removed by heavy machines, often over a very large area, to expose the ore beneath. In cases where it is not economical to remove the surface, shafts are dug into the earth, with side tunnels to follow the layer of ore.

Refining

13   The mined ore is crushed and sorted. The best grades of ore contain over 60% iron. Lesser grades are treated, or refined, to remove various contaminants before the ore is shipped to the blast furnace. Collectively, these refining methods are called beneficiation and include further crushing, washing with water to float sand and clay away, magnetic separation, pelletizing, and sintering. As more of the world's known supply of high iron content ore is depleted, these refining techniques have become increasingly important.

14   The refined ore is then loaded on trains or ships and transported to the blast furnace site.

The Manufacturing Process

Charging the blast furnace

15   After processing, the ore is blended with other ore and goes to the blast furnace. A blast furnace is a tower-shaped structure, made of steel, and lined with refractory, or heat-resistant bricks. The mixture of raw material, or charge, enters at the top of the blast furnace. At the bottom of the furnace, very hot air is blown, or blasted, in through nozzles called tuyéres. The coke burns in the presence of the hot air. The oxygen in the air reacts with the carbon in the coke to form carbon monoxide. The carbon monoxide then reacts with the iron ore to form carbon dioxide and pure iron.

Separating the iron from the slag

16   The melted iron sinks to the bottom of the furnace. The limestone combines with the rock and other impurities in the ore to form a slag which is lighter than the iron and floats on

top. As the volume of the charge is reduced, more is continually added at the top of the furnace. The iron and slag are drawn off separately from the bottom of the furnace. The melted iron might go to a further alloying process, or might be cast into ingots called pigs. The slag is carried away for disposal.

Treating the gases

17 The hot gases produced in the chemical reactions are drawn off at the top and routed to a gas cleaning plant where they are cleaned, or scrubbed, and sent back into the furnace; the remaining carbon monoxide, in particular, is useful to the chemical reactions going on within the furnace.

18 A blast furnace normally runs day and night for several years. Eventually the brick lining begins to crumble, and the furnace is then shut down for maintenance.

Quality Control

19 The blast furnace operation is highly instrumented and is monitored continuously. Times and temperatures are checked and recorded. The chemical content of the iron ores received from the various mines are checked, and the ore is blended with other iron ore to achieve the desired charge. Samples are taken from each pour and checked for chemical content and mechanical properties such as strength and hardness.

Byproducts/Waste

20 There are a great many possible environmental effects from the iron industry. The first and most obvious is the process of open pit mining. Huge tracts of land are stripped to bare rock. Today, depleted mining sites are commonly used as landfills, then covered over and landscaped. Some of these landfills themselves become environmental problems, since in the recent past, some were used for the disposal of highly toxic substances which leached into soil and water.

21 The process of extracting iron from ore produces great quantities of poisonous and corrosive gases. In practice, these gases are scrubbed and recycled. Inevitably, however, some small amounts of toxic gases escape to the atmosphere.

22 A byproduct of iron purification is slag, which is produced in huge amounts. This material is largely inert, but must still be disposed of in landfills.

23 Ironmaking uses up huge amounts of coal. The coal is not used directly, but is first reduced to coke which consists of almost pure carbon. The many chemical byproducts of coking are almost all toxic, but they are also commercially useful. These products include ammonia,

which is used in a vast number of products; phenol, which is used to make plastics, cutting oils, and antiseptics; cresols, which go into herbicides, pesticides, pharmaceuticals, and photographic chemicals; and toluene, which is an ingredient in many complex chemical products such as solvents and explosives.

24 Scrap iron and steel—in the form of old cars, appliances and even entire steel-girdered buildings—are also an environmental concern. Most of this material is recycled, however, since steel scrap is an essential resource in steelmaking. Scrap which isn't recycled eventually turns into iron oxide, or rust, and returns to the ground.

The Future

25 On the surface, the future of iron production—especially in the United States—appears troubled. Reserves of high-quality ore have become considerably depleted in areas where it can be economically extracted. Many long-time steel mills have closed.

26 However, these appearances are deceiving. New ore-enrichment techniques have made the use of lower-grade ore much more attractive, and there is a vast supply of that ore. Many steel plants have closed in recent decades, but this is largely because fewer are needed. The efficiency of blast furnaces alone has improved remarkably. At the beginning of this century, the largest blast furnace in the United States produced 644 tons of pig iron a day. It is believed that soon the possible production of a single furnace will reach 4,000 tons per day. Since many of these more modern plants have been built overseas, it has actually become more economical in some cases to ship steel across the ocean than to produce it in older U.S. plants.

三、分析与讲解

铁是最重要的金属之一。铁的发现、冶炼和使用直接将人类从青铜器时代带入铁器时代。没有铁以及铁进一步冶炼得到的钢，便没有工业革命，没有现代化大生产，也将没有今天的电子时代、信息时代。

这篇文章简单介绍了铁，包括人类使用铁的历史、铁冶炼工艺的演变、冶炼铁矿石需要的原材料、铁矿的开采和处理、炼铁流程及副产品和将来的发展趋势等。文章带有科普性质，因此虽然文章中有一些冶金术语，但整体难度不大，翻译起来也比较容易。不过还是存在一些难点，兹分析如下：

第 1 段前小标题"Background"不能简单译成"背景"，因为接下来的文字并不是提供背景信息。需要根据小节的内容将其译为"引言"或"引子"。

科技翻译也好，其他类型的翻译也罢，标题和小标题的翻译都不能简单地根据原文标题或小标题的意思来定，而要根据标题所在整篇文章或小标题所在小节的内容来决定，有时甚至要重拟标题。

第 1 段：(1)第二句中的 construction 不容易翻译。英汉词典对 construction 的解释是"建造，构筑，建设；建造物，构筑物，结构，建筑；建造术，构筑术；意义，解释"，但这些释义都不合适。英英词典中 construction 的释义是 a. The act or process of constructing; b. The art, trade, or work of building; c. A structure, such as a building, framework, or model; d. The way in which something is built or put together; e. The interpretation or explanation given to an expression or a statement。这里的 construction 事实上外延很广，包括人类制造的每一样物件、建造的每一栋建筑物。简单起见，就译成"物件"；或者复杂一点：(制造的每一样)物件、(建造的每一栋)建筑。(2) the oldest metals 字面意思是"最古老的金属"，但金属无所谓"古老"与否，所以这里最好译成"使用历史最久的金属"。

第 2 段：(1)第 1 句按照原文的结构可以翻译成"纯铁是一种软的、灰白色的金属"，意思没问题，但不够地道——汉语中"是"字句用得较少，该句式带有西化的意味，所以一般来说英语原文的"be+表语"结构，译成汉语时都不要译成"是"字结构。同英语一样，科技汉语表达讲究简洁，所以此句不妨译作"作为一种金属，纯铁质柔，表面呈灰白色"。(2) Iron oxides are the most common 这句最好也不译作"是"字句，而译为"铁氧化物最为常见"。

第 3 段：最后一句结构有点复杂。前半部分 Pig iron has only limited uses 直译成"生铁的用途有限"也可以，但不如采用正说反译，即用否定句翻译肯定句，译成"生铁用途不广"。后半部分主句中带定语从句，从句中又有"介词+动名词结构"以及不定式目的状语，翻译时需要在准确理解意思的基础上调整句序，先译手段 by further reducing... and adding...，再译目的 to give the steel specific properties 和 is converted into various steel alloys.

科技英语以客观陈述为主，一般不带感情色彩，体现了内容上的科学性和客观性，大量使用非谓语动词(过去分词、现在分词、动名词和动词不定式)、被动语态及带有后置定语从句和介词、形容词、副词结构的长句。这些结构的使用一方面使句子结构严谨，逻辑性强，但另一方面也加大了理解和翻译的难度。遇到这类句子时，要仔细研读原文，分析句子结构，理清脉络，明确成分之间的关系，确保整体上理解原文后，才能正确进行翻译。

第 4 段：(1)不要将 its production was a closely guarded secret 译成"铁的生产是一个细心守护的秘密"，那样太别扭；可以说"铁生产工艺被当作秘密细心守护"。(2)科技翻译力求简洁，tools made from it 直译是"用铁所做的工具"，也就是"铁制工具"。(3) The chemical nature of... were not understood 是被动语态，翻译时最好译成主动句，由泛指"人们"作主语：人们并不了解铁提炼过程中发生的变化背后的化学原理。(4) heirloom 意思是"祖传遗物；传家宝"，这里的 heirloom swords 就可以简单译为"传下来的宝剑"。前面的

produced 要译为"生产出了"而不是"生产了"。(5) but the processes never emerged into the rest of the world 中的 the processes 用的是复数，所以不仅仅指"铁的生产工艺"，而应该译为"相关工艺"。(6) For centuries 中的 centuries 非确指，不能译为"几个世纪"或"几百年来"，因为有可能多于"几个世纪"。可以译成"在相当长的时间里"。

将英语被动句译成泛指词语（"人们""有人""大众"等）作主语的汉语主动句是常用的英汉翻译技巧之一，这种技巧也广泛用于科技翻译。例如：

Rubber is found to be a good isolating material.
人们发现，橡胶是一种良好的绝缘材料。
With the development of network technology, information can be transmitted to wherever it is needed.
随着网络技术的发展，人们可以把信息传播到任何需要它的地方。
It is believed that soon the possible production of a single furnace will reach 4,000 tons per day.
人们相信，很快单座高炉日产就能达到4000吨。

第5段：(1) will hold a cutting edge 就是"将有锋利的边缘"的意思，但不能将 will 译为"将"，因为不是所有的铸铁产品都有锋利的边缘，只是在有必要的情况下，可以如此翻译。(2) iron workers 不要想当然地译为"铁匠"，英文中对应"铁匠"的词语是 ironsmith 或 blacksmith，这里的 iron workers 是"制铁工人"。(3) 注意 Entire nations 中 nation 用的是复数，所以并不是"全国"，而是"各国"。

第6段：(1) took out a patent 不仅"申请"而且"获批了"。(2) 一般来说，blast furnace 需要译为"高炉"，但 Huntsman used a blast furnace 中的 blast furnace 并不是我们今天意义上的"高炉"，这里可译为"鼓风炉"。(3) a side-effect 是一种幽默的说法，这里当然不能译为"副作用"，不妨将主句译为"亨茨曼顺便发明了现代冶金术"。

第8、9段间小标题：查词典，refining 有"精致，精炼，提炼，提纯，精选，清选，加工，净化，纯化，匀料，加细，清扫，清洗"等释义，但阅读这一小节的文字，可以发现这几段并不是讨论铁矿石精炼，而是叙述如何处理铁矿石，所以可以将 refining 译为"处理"。

第10段：(1) 一般来说，科技翻译需要使用比较正式的语言，例如 Called refining "puddling", this was highly skilled work, but was also hot, strenuous, and dangerous 虽然可以译为"这种工作被称为'搅炼'，要求较高的技术，但又热又累又危险"，但明显不如"这种工作被称为'搅炼'，技术性很强，但工作环境酷热，劳动强度大，危险性高"。(2) hearty constitution 是"强健的体格"。

第12段：第1句中的 removed 不好翻译，考虑到语境，可以译为"剥去"；the surface of the ground 比较抽象，这里可以具体化译为"地表的土石"。

第14、15段间小标题：Charging the blast furnace 如果简单译为"给高炉装料"，有点不太像标题，不妨译为更具技术结构的词组："装料上炉"。

第15段：第2句是个判断句，但正如前面所说，汉语中虽然有"是"字句这样的判断句，但使用频率不高，故此句可译为"高炉呈塔形，由钢制造，内衬耐火材料或耐热砖"。

第15、16段间小标题：Separating the iron from the slag 最好直接按原结构译成"将铁与渣分开"而不是"渣铁分离"，因为"渣铁分离"带有"渣和铁主动分离"的意思，而将渣和铁分开是人为干预的结果。

第19段：(1)第1句 The blast furnace operation is highly instrumented 可以译为"高炉装有很多仪器仪表"，也可以译为"高炉生产需要很多仪器仪表"。(2)最后一句中的 each pour 当然是 pour of melted iron，也就是"每次出铁"。

四、参考译文

<h3 style="text-align:center">铁</h3>

引言

1 铁是地球上最常见的元素之一，几乎人类(制造/创造/生产)的每一个物件都或多或少包含铁。铁也是使用历史最久的金属之一，至少在3500年前，人类已首次将铁加工成有用的物品或装饰品。

2 作为一种金属，纯铁质柔，表面呈灰白色。虽然铁作为一种元素铁很常见，但纯铁在自然界中却几乎未见。已知自然界中仅见的纯铁来自陨石。大部分铁是以与其他元素结合的形式存在于矿物中的，其中铁氧化物最为常见。地球表层含铁量最高的那些矿物称为铁矿，可以进行商业开采。

3 通过几种工艺，可将铁矿石转换为不同种类的铁。最常见的工艺是高炉，高炉生产的生铁含铁92%—94%，含碳3%—5%，此外还含有少量的其他成分。生铁用途不广，大部分生铁被送进钢厂，通过进一步降低碳含量以及添加像锰和镍等元素，改善钢的特性，生产钢合金。

历史

4 历史学家相信，埃及人最早使用少量的铁，时间大约在五六千年前。很明显，他们使用的铁来自陨石。被认为可以当作最早进行铁开采和冶炼的证据，是在古代赫梯文明，也就是今天的土耳其发现的。因为铁用来制造武器和工具较其他已知金属优越得多，所以铁生产工艺被当作秘密细心守护。然而，生产铁的基本技术并不复杂，因此铁的使用逐渐传播开来。尽管与其他材料相比，铁用途多样，但铁也有劣势。取决于铁矿石所来自的地区及铁的提炼方法，铁制工具的质量非常参差不齐。人们并不了解铁提炼过程中发生的变化背后的化学原理，尤其是碳对于铁的硬度的重要影响。世界各地的炼铁工艺差别很大。例如，有证据表明，中国人很早就能将铁熔化，铸造铁制器具，而日本人多个世纪前传下来的宝剑证明了，他们使用少量的钢生产出了令人惊奇的产品。中东和印度也有类似的突破，但相关工艺却没有传到世界其他地区。在相当长的时间里，欧洲人始终不知道如何将铁加热到熔点。为了生产铁，他们将铁矿石和木柴放在陶制(黏土)衬里的炉子里，燃烧后，铁同周围岩石分离，但却不会充分熔化，而是形成坚硬的渣层，通过锤子捶打去除。反复加热和捶打让氧和铁氧化物混合产生铁，同时去除金属中的碳。得到的是近乎纯铁，易于用锤子和钳子加工，但纯铁太软，边缘不够坚硬(强度不够)。因为这种金属能够用锤子加工锻打，所以被称为锻铁(熟铁)。

5　从东方带回欧洲的工具和武器是由熔化后铸造成形的铁制成的。铸铁含碳量比锻铁高，因此更硬，可以有锋利的边缘，但比锻铁脆。欧洲的制铁工人知道东方人制造的铁更好，但不了解制造更坚硬铁制品的工艺。各国都在努力寻找这种工艺。

6　已知欧洲在铸铁生产上的第一次突破，直到1740年才出现，很快，更具实用价值的钢得以首次生产。那一年，本杰明·亨茨曼申请获批了一项熔化材料生产钢弹簧的专利，用于钟表制作。接下来的大约20年里，亨茨曼发明的工艺得到了广泛的使用。亨茨曼在黏土坩埚里通过鼓风熔化熟铁，然后在熔化的铁水里加上适量的纯木炭，生成的合金铸成弹簧，不仅强度大，而且弹性好。因为亨茨曼最初的兴趣只是制作更精密的时钟，他的坩埚钢直接助推了航海用天文钟的发明，因为天文钟能够让水手准确地确定所在的东西方位，使得航行全球成为可能。至于亨茨曼顺便发明了现代冶金术，明显是他没有注意到的。

原材料

7　高炉生产生铁用到的原材料有铁矿石、焦炭、烧结料和石灰石。铁矿石主要是铁氧化物，包括磁铁矿、赤铁矿、褐铁矿和其矿石。这些矿石的铁含量从70%到20%甚至更少不等。焦炭是将煤加热直至变成接近纯碳制成的。烧结料是用品位较低的铁矿石破碎成细粉，同焦炭、石灰一起焙烧，去除矿石中大量杂质而制得。石灰石存在于自然界中，是碳酸钙的来源。

8　生产各种类型的钢时，有时也会将其他金属加入铁中，如铬、镍、锰、钼和钨。

铁矿石的开采和处理

9　铁矿石用于高炉之前，须从地下开采出来，经过部分处理，去除大部分杂质。

10　历史上，铁最初是用热风法生产的，后来采用的是无烟煤炉。两种方法生产铁时，基本动作都是工人不停地搅拌小炉的生铁和炭屑，直至铁渣分离。这种工作被称为"搅练"，技术性很强，但工作环境酷热，劳动强度大，危险性高。搅练需要丰富的经验和强健的体格。搅练工人独立，自豪，薪水很高。

11　搅练工人于1858年在匹兹堡成立了钢铁行业的第一个工会"火神之子"。1876年，该协会和另外三个劳工协会合并，成立了钢铁工人联合协会。1892年，在荷姆斯特德罢工中，钢铁工人联合协会被安德鲁·卡内基击败，从而分崩离析，钢铁产业基本进入无序状态，直至20世纪30年代。

开采

12　世界上大部分铁矿石都是露天开采，重型机械剥去地表的土石(通常是很大一块区域)，裸露出下面的矿床。如果露天开采不经济，则顺着矿层，朝地下挖掘矿井，开凿侧向巷道。

处理

13　开采出来的矿石需要破碎、分选。最高品位的矿石含铁可达60%以上。较低品位的矿石在送进高炉前需要处理或加工，去除各种杂质。这些处理方法总称为选矿，包括进一步的破碎、水洗去除沙土、磁选、造球和烧结。随着世界上已知高含量矿石储量越来越少，这些处理技术已变得越来越重要。

14　处理过的矿石被装进火车或轮船，运到高炉所在地。

生产过程

装料上炉

15　经过处理后，矿石和其他原材料混合，被送入高炉。高炉呈塔形，由钢制造，内衬耐火材料或耐热砖。混合的原材料，即炉料，从高炉顶部入炉。在高炉底部，通过被称为风口的喷嘴，灼热空气被吹进高炉。焦炭在热空气中燃烧。空气中的氧与焦炭中的炭发生反应生成一氧化碳，一氧化碳再与铁矿石发生反应，生成二氧化碳和纯铁。

将铁与渣分开

16　熔化的铁沉到炉底。石灰石与矿石中的岩石和其他杂质结合，形成炉渣，炉渣比铁轻，会浮在铁水上面。随着炉内炉料的减少，更多炉料持续不断地从炉顶装入，铁、渣分别从炉底排出。铁水或者经过进一步合金化处理，或者铸成铁锭，称为铸块。炉渣被运走，以便处理。

炉气处理

17 化学反应产生的炉气从炉顶排出，被送进气体净化工厂，净化或洗涤后送回高炉。特别是，剩下的一氧化碳对于炉内发生的化学反应特别有用。

18 高炉通常在几年时间里昼夜运转，最终，砖衬开始剥落，高炉需要停产检修。

质量控制

19 高炉生产需要很多仪器仪表，进行不间断监测。时间和温度需要核对并记录，来自不同铁矿铁矿石的化学成分需要测定，为了得到理想的炉料，需要将不同的铁矿石按比例混合。每次出铁，都需要抽样检查铁的成分及强度、硬度等机械特性。

副产品/废弃物

20 铁的生产会带来很多可能的环境效应。排在首位也是最明显的是铁矿的露天开采。大片土地被剥离，裸露出岩石。今天，资源枯竭的煤矿常被用作垃圾填埋场，然后再盖上土，进行景观改造。其中一些垃圾填埋场自身造成了环境问题，因为近年来，有些填埋场被用于处理高毒性物质，它们渗进了土壤和水中。

21 从铁矿石中提取铁会产生大量的有毒的、腐蚀性气体。生产实践中，这些气体会被净化再利用。但不可避免地，会有少量有毒气体逃逸到大气中。

22 铁冶炼带来的一个副产品是大量的炉渣，大部分炉渣性质不活泼，但也必须填埋处理。

23 炼铁消耗大量的煤。煤不是直接用于炼铁，而是先被还原成焦炭，焦炭几乎就是纯碳。炼焦产生的很多化学副产品虽然几乎都有毒性，但也都有商业用途。这些副产品包括可用于多种产品的氨，可用于制造塑料、切削油和防腐剂的酚，可用于除草剂、杀虫剂、药品、照相化学品的甲酚，以及很多复杂化学品，如溶剂、炸药的组成成分甲苯。

24 来自旧汽车、器械甚至整个钢结构建筑的废钢铁也会造成环境问题。但是，因为废钢是炼钢的基本原材料，大部分废钢铁都会被循环利用，没有被循环利用的废钢铁最终会转变成氧化铁即铁锈，回到自然界(土壤)。

未来发展

25 表面看来，钢铁生产的未来，特别是美国钢铁生产的未来，看起来不容乐观。在很多地区，可以经济地开采的高品位的矿石储量急剧减少，很多历史较久的钢厂关门停业。

26 然而，这些表象带有欺骗性。新研发的矿石富集技术让低品位矿石的使用变得更有吸引力，而低品位矿石的供应量很大。确实，过去几十年来，很多钢厂关闭了，但这主要是因为不需要那么多钢厂。仅高炉的生产效率就提高了很多。本世纪初，美国最大的高炉日产生铁 644 吨，而人们相信，很快单座高炉日产就能达到 4000 吨。因为很多更现代化的工厂建在海外，所以实际上在某些情况下，从海外跨洋将钢铁运到美国比在美国的老钢厂里生产钢铁还要更经济。

五、专题讨论

科技翻译中的选词

翻译任何材料，选词都是必不可少的一步，通常也是表达的第一步。词语选错了，不仅影响原文意义的准确传达，也会影响译文表达的流畅。科技翻译以准确传达科技信息为主要目标，词语选错了，必然会影响科技信息的准确再现，甚至会直接影响科技产品的制造、科研仪器的使用或科学实验的开展。

英汉科技翻译也好，汉英科技翻译也罢，要做好选词工作，首先要对科技英语中的用词特点有清晰的了解。

（一）科技英语的用词特点

科技英语中使用的词汇大致可分为三类：普通词汇、半专业词汇和专业词汇。

普通词汇很好理解，就是我们在日常工作、生活中使用的词汇。例如：

Iron is one of the most important elements on Earth. Nearly every construction of man contains at least a little iron.

在这句里，几乎所有的词汇都是普通词汇，也就是大部分学过英语的人都能够理解的词汇。

半专业词汇涉及一定的专业学科知识，但受过通识教育的人基本上能够理解，例如 cast iron（铸铁）、iron ore（铁矿石）、coke（焦炭）。半专业词汇与普通词汇没有明确的界限，判断一个词语是普通词汇还是半专业词汇，只能说是仁者见仁、智者见智。比如上述例句中的 element（元素），可以说是普通词汇，也可以说是半专业词汇。

专业词汇也可以称为专业术语，涉及较深的专业知识，非专业人士有时很难理解，比如 sinter（烧结料）、cleavage（解理）、mudgun（泥炮）、bustle pipe（热风围管）、reductant

(还原剂)、matte(锍)、counter-current reactor(逆流反应器)、variable(变量)等。

半专业词汇、专业词汇除专门为专业表达创造的词语(如 titanium、magnetite、chiton、bauxite、gangue、joule、calorie 等)外,还有几个来源:

一是普通词汇转用。如:

词语	普通用法	专业用法	词语	普通用法	专业用法
bell	钟	料钟	section	部分	型材
grade	年级	品位;钢种	charge	收费,指控	炉料;装料
bath	浴缸;沐浴	熔池	heat	热量;加热	炉次

二是通过添加前缀、后缀等派生构词法产生。如:

convert + er = converter(转炉)

de + sulphur + ization = desulphurization(脱硫)

transform + er = transformer(变压器)

pellet + ize = pelletize(造球,使成团)

electro + metallurgy = electrometallurgy(电冶金学)

三是将两个或两个以上的词放在一起组成合成词或词组。如:

up + take = uptake(上升管)

iron + works = ironworks(铁厂)

tap + hole = taphole(出铁口;出渣口)

blast + furnace = blast furnace(高炉)

carbon + dioxide = carbon dioxide(二氧化碳)

一般说来,科普文章或著作中使用普通词汇、半专业词语较多,文章或著作的专业性越强,所使用词汇的专业性也越强,学术论文、专著等更是大量使用专业词汇和术语。

科技英语以完整、准确、高效传达科技信息为主要目的,自然地,科技英语以平实、质朴、简洁、严谨为风格特征,表现在词语的使用上:

第一,多用意义明确的词语,少用意义含混的词语。

第二,多用名词、动词,少用形容词、副词(科技英语中为数不多的形容词、副词也以描述性形容词、副词为主,少用评价性形容词、副词,特别是带有感情色彩的形容词、副词)。

第三,多用偏正式的规范词语,少用口语词汇,罕用或不用俚语。

第四,多用数字、公式、符号等,不用这些数字、公式、符号等对应的英语表达式。

第五,为了更有效率地传递科技信息,科技英语中还大量使用缩略语。如:BF(blast

furnace)、BOS（basic oxygen steelmaking）、EAF（electric arc furnace）、EBT（eccentric bottom tap-hole）、DRI（direct reduced iron）、HM（hot metal）、CSR（coke strength after reaction with CO_2）等。

（二）科技翻译选词的原则

虽然科技英语多用意义精确的词语，但这并不意味着科技英语翻译中选词就是一件容易的事，这是因为：同样一个词在不同的学科、不同的语境可能会有不同的意思，可能需要与不同的词搭配。特别是，用在科技英语中的普通词汇及从普通词汇转用的半专业或专业词汇，大多具有多个意思，如何在多个意思中取舍，常常是个问题。例如本课课文中的小标题之一"The Ore Extraction and Refining Process"中，refine一词在词典中有"精致，精炼，提炼，提纯，精选，清选，加工，净化，纯化，匀料，加细，清扫，清洗"等释义，但最后却需要根据小节内容译为"处理"。

还比如：

To produce iron, they slowly burned iron ore with wood in a clay-lined oven.

句中，wood一词有"木材，木头，木料，树林"等意思，考虑到这里的wood是用来燃烧的，所以译为"木柴"更合适。

专业词汇的翻译也会成为问题，比如carbon这个词，出现在carbon bond、carbon chain、carbon monoxide、carbon CP、carbon AR这些词组中时必须译为"碳"（碳键，碳链，一氧化碳，化学纯碳，分析用碳），而出现在carbon black、active carbon、carbon brick、carbon electrode、carbon blanket等词语中时需要译为"炭"（炭黑，活性炭，炭砖，炭电极，炭毯）[①]

一般说来，科技翻译中选词须遵循四个原则：

1. 根据文本或句子所涉及的学科来进行选择。如前所述，同一个词语出现在不同学科、不同专业中，其含义及汉语表达都可能不同，译者必须结合学科和专业来正确地理解词语并选择合适的译法。例如：

Power can be transmitted over a long distance.

电力可以输送到很远的地方。

This is a 20 power binoculars microscope.

这是一架20倍的双镜头显微镜。

Wheat has not done well, though it is jointing now.

小麦长势不太好，尽管现在正在拔节。

The X-ray and CT scan can clearly show the anatomic structure and pathological

---

① 魏寿昆. 选用"碳""炭"的简明方法[J]. 科技术语研究，2006(3)：5.

feature of facet joint, and offer reliable radiological evidence to clinic.

X射线及CT扫描能够清楚地显示椎小关节的解剖结构及病理改变,为临床提供了可靠的影像学依据。

The joints which connect two mechanic components of mechanisms can make sure the machines move relatively.

连接机构两个构件的运动副能保证构件间可以有一定的相对运动。

Analytical methods for dynamics model of machine tool structure with joints were studied in this paper.

本文研究了带结合部机床结构的动力学特性的系统建模解析方法。

2. 根据词的搭配进行选择。英、汉两种语言中的词,各有一套自己的搭配系统,主要表现在形容词与名词、动词与名词、动词与副词等的搭配关系上。科技翻译时,须在正确理解原文的基础上,根据汉语的习惯搭配来处理原文中的搭配。例如:

British rail engineers at its Railway Technical Center at Derby have trimmed weight and drag substantially.

位于德比郡的铁路技术中心的英国铁路工程师大大地减轻了列车的重量,增加了牵引力。

此句中的 trim 一词和两个不同的宾语 weight 和 drag 搭配,也因此具有了两种截然不同的含义,可分别译为"减轻"和"增加"。

3. 根据词性来进行选择。一个英语词汇往往兼有不同词性,分属于几种不同的词类。在大多数场合下,同一个词因为词性不同会有词义的不同。一词多类多义时,要选择词义,译者首先要确定这个词在句中属于哪一种词类,起什么作用,然后再根据词类选择一个适当的词义。例如:

Like charges repel, unlike charges attract.
同性电荷相斥,异性电荷相吸。(形容词)
In the sunbeam passing through the window there are fine grains of dust shining like gold.
在射入窗内的阳光里,细微的尘埃像金子一般在闪闪发亮。(介词)
It is the atoms that make up iron, water, oxygen and the like.
正是原子构成了铁、水、氧等类物质。(名词)
He likes making experiments in chemistry.

他喜欢做化学实验。（动词）

Waves in water move like the waveform moves along a rope.

波在水中移动就像波形沿着绳子移动一样。（连词）

4. 根据上下文语境来进行选择。科技文章具有较强的科学性和逻辑性，选择词义时，应该做到词不离句，句不离文。为求得正确理解，要文中求句，句中求词。例如：

Rubber is not hard, it gives way to pressure.

橡胶不硬，受压就会变形。

The overpass might give way during an earthquake.

这座立交桥在地震时可能会倒塌。

I believe old farming methods should give way to improved modern ones.

我认为旧的耕作方式应给改良了的现代化的耕作方式让路。

这四个原则对科技英语中普通词汇、半专业词汇和专业词汇的翻译都适用。不过对于半专业词汇和专业词汇，还有一个"统一规范"的翻译原则必须遵守，关于这点以及相关内容，将在第三课"专题讨论科技术语的翻译"（57 页）中进行详细分析。

◎ 练习二　将下面的句子翻译成汉语。

1. The beginning of the extraction of iron from its ores dates back to prehistoric times. In early times, iron ore was heated in a charcoal fire (doubtless by chance at first). When the fire went out, a piece of solid iron like a sponge was left. Spongy iron could be hammered into shape to make tools and weapons.

2. Worldwide, the iron and steel industry is one of the most significant and, in terms of tradition, one of the oldest sectors of industry. As early as 3,000 years ago, iron was serving as a basis of human culture and civilization.

3. The name iron is derived from the Etruscan word "aiser" meaning "the gods" because the earliest iron was obtained from meteorites and meteorites fall from the sky.

4. Combined with varying (but tiny) amounts of carbon, iron makes a much stronger material called steel, used in a huge range of human-made objects, from cutlery to warships, skyscrapers, and space rockets.

5. Iron's major drawback as a construction material is that it reacts with moist air (in a process called corrosion) to form the flaky, reddish-brown oxide we call rust.

6. Elements can be classified based on their physical states (States of Matter), e.g., gas,

solid or liquid. Iron is a solid. Iron is classified as a "Transition Metal". Elements classified as Transition Metals are generally described as ductile, malleable, and able to conduct electricity and heat.

7. The Bessemer process was the first inexpensive industrial process for the mass-production of steel from molten pig iron. It involved blowing air through molten pig iron to burn off carbon, and so to produce mild steel.

8. The beginning of the extraction of iron from its ores dates back to prehistoric times. In early times, iron ore was heated in a charcoal fire (doubtless by chance at first). When the fire went out, a piece of solid iron like a sponge was left. The spongy iron could be hammered into shape to make tools and weapons. Our metallurgical forefathers found that when they blew or fanned the flames, the fire became hotter and the iron was produced more rapidly, so bellows were used to increase the supply of air.

9. Before the Industrial Revolution, steel was an expensive material, produced in only small quantities for such articles as swords and springs, while structural components were made of cast iron or wrought iron.

10. Warfare hugely increases steel's worldwide production. The 20th century's two world wars had huge consequences for steelmaking. Like other heavy industries, steelmaking was nationalized in many countries due to demands for military equipment.

第三课 磁 铁 矿

> "矿产资源是经济社会发展的重要物质基础……"
> ——习近平《给山东省地矿局第六地质大队全体地质工作者的回信》①

一、科技术语

banded iron formation Also known as "banded ironstone formations" or BIFs, distinctive units of sedimentary rock, almost always of Precambrian age, that consist of repeated, thin layers (a few millimeters to a few centimeters in thickness) of silver to black iron oxides, either magnetite or hematite, alternating with bands of iron-poor shales and cherts, often red in color, of similar thickness, and containing microbands of iron oxides.(条带状含铁建造)

biomagnetism An interdisciplinary subject, studying connection and affection between magnetic field and organisms.(生物磁学)

caltech California Institute of Technology, a world-renowned science and engineering Institute.(加州理工学院)

chemical formula A representation of a substance using symbols for its constituent elements.(化学分子式)

chiton Small to large marine molluscs.(石鳖)

cleavage The splitting or tendency to split of a crystallized substance along definite crystalline planes, yielding smooth surfaces.(晶体解理)

crystal habit (Chemistry) the external shape of a crystal.(晶体形态)

crystal system In crystallography, the terms crystal system, crystal family, and lattice system each refer to one of several classes of space groups, lattices, point groups, or crystals. Informally, two crystals tend to be in the same crystal system if they have similar symmetries,

---

① 习近平. 给山东省地矿局第六地质大队全体地质工作者的回信[N]. 光明日报, 2022-10-05(1).

though there are many exceptions to this.(晶系)

dunite An ultrabasic igneous rock consisting mainly of olivine. (纯橄榄岩)

fracture The characteristic manner in which a mineral breaks; the characteristic appearance of the surface of a broken mineral.(断口)

igneous Formed by solidification from a molten state.(火成的)

ilmenite A weakly magnetic titanium-iron oxide mineral which is iron-black or steel-gray. (钛铁矿)

magma The molten rock material under the earth's crust, from which igneous rock is formed by cooling. (岩浆)

magnetoreception A sense which allows an organism to detect a magnetic field to perceive direction, altitude or location. This sense has been proposed to explain animal navigation in vertebrates and insects, and as a method for animals to develop regional maps. (地磁感受)

magnetospirillum Magnetotactic bacteria . (趋磁螺旋菌)

metamorphic rock Rocks formed by the action of heat or pressure.(变质岩)

paleoecologist A scientist who studies the ecology of the geologic past.(古生态学家)

paleomagnetism Study of the intensity and orientation of the earth's magnetic field as preserved in the magnetic orientation of certain minerals found in rocks formed throughout geologic time.(古磁力学)

peridotite A dense, coarse-grained igneous rock, consisting mostly of the minerals olivine and pyroxene.(橄榄岩)

radula A horny tooth-bearing strip on the tongue of molluscs that is used for rasping food. (软体动物用来锉磨食物的齿舌)

sedimentary Of or relating to rocks formed by the deposition of sediment.(沉积岩)

serpentinization A hydrothermal process by which magnesium-rich silicate minerals are converted into or replaced by serpentine minerals. (蛇纹石化)

solid solution A homogeneous crystalline structure in which one or more types of atoms or molecules may be partly substituted for the original atoms and molecules without changing the structure.(固溶体).

streak The color of the fine powder produced when a mineral is rubbed against a hard surface. Used as a distinguishing characteristic.(条痕)

ulvospinel A mineral of the subclass of multiple oxides, Fe_2TiO_4, which crystallizes in the isometric system and whose structure is the inverted form of the spinel structure. Ulvospinel is a component of complex titanium-magnetite ores.(钛尖晶石)

二、英语原文

Magnetite

1 Category: Mineral

Chemical formula: iron(Ⅱ, Ⅲ) oxide, Fe_3O_4

Color: Black, greyish

Crystal habit: Octahedral, fine granular to massive

Crystal system: Isometric

Cleavage: Indistinct

Fracture: Uneven

Mohs Scale hardness: 5.5-6.5

Luster: Metallic

Refractive index: Opaque

Streak: Black

Specific gravity: 5.17-5.18

2 Magnetite is a ferromagnetic mineral with the chemical formula Fe_3O_4 and the common chemical name ferrous-ferric oxide, which indicates the mineral comprises both a ferrous component, FeO (wüstite), and a ferric component, Fe_2O_3 (hematite). Magnetite is one of several types of iron oxide and its official (IUPAC) name is iron(Ⅱ, Ⅲ) oxide. It is a member of the spinel group of minerals, which crystallize in cubic and octahedral patterns, and its crystals are black and opaque. The most magnetic of all naturally occurring minerals on Earth, magnetite occurs in some places as naturally magnetized stone called lodestone and was used as an early form of magnetic compass. Magnetite dissolves slowly in hydrochloric acid.

3 Magnetite mineral is valuable as an iron-bearing ore. In addition, as it carries the dominant magnetic signature in rocks and it tends to lock in the magnetic pattern it carried as it was last hardening, magnetite has played a critical role in understanding plate tectonics. Changes in the oxygen content of the Earth's atmosphere can be inferred by studying sedimentary rocks containing magnetite. Moreover, interactions between magnetite and other oxide minerals have been studied to determine the oxidizing conditions and evolution of magmas over geological history.

4 Small grains of magnetite occur in almost all igneous rocks and metamorphic rocks.

Magnetite also occurs in many sedimentary rocks, including banded iron formations. In many igneous rocks, magnetite-rich and limonite (a titanium iron oxide)-rich grains occur that precipitated together from magma.

Occurrence

5 Magnetite occurs in many sedimentary rocks, and huge deposits have been found in banded iron formations. In addition, this mineral (especially in the form of small grains) occurs in almost all igneous and metamorphic rocks. Magnetite is also produced from peridotites and dunites by serpentinization.

6 Magnetite is sometimes found in large quantities in beach sand. It is carried to the beach by the erosive action of rivers and is concentrated by waves and currents. Such mineral sands (also called iron sands or black sands) are found in various places, including beaches in California and the west coast of New Zealand. In June 2005, an exploration company (Candero Resources) discovered a vast deposit of magnetite-bearing sand dunes in Peru, where the highest dune is more than 2,000 meters (m) above the desert floor. The dune field covers 250 square kilometers (km^2), and ten percent of the sand is magnetite.

7 Large deposits of magnetite have been found in Kiruna, Sweden, and the Pilbara region of Western Australia. Additional deposits occur in Norway, Germany, Italy, Switzerland, South Africa, India, and Mexico. In the United States, it is found in the states of New York (Adirondack region), New Jersey, Pennsylvania, North Carolina, Virginia, New Mexico, Colorado, Utah, and Oregon.

Biological Occurrences

8 Crystals of magnetite have been found in some bacteria (such as Magnetospirillum magnetotacticum) and in the brains of bees, termites, some birds (including pigeons), and humans. These crystals are thought to be involved in magnetoreception—the ability to sense the polarity or inclination of the Earth's magnetic field—and to aid in navigation. Also, chitons have teeth made of magnetite on their radula, making them unique among animals. This means they have an exceptionally abrasive tongue with which to scrape food from rocks. The study of biomagnetism began with the discoveries of Caltech paleoecologist Heinz Lowenstam in the 1960s.

Laboratory Preparation

9 Magnetite can be prepared in the laboratory as a ferrofluid using the Massart Method. It

involves mixing iron(Ⅱ) chloride and iron(Ⅲ) chloride in the presence of sodium hydroxide.

Characteristics

10　This mineral is the most magnetic of all known naturally occurring minerals. Its Curie temperature is about 580℃. Chemically, it dissolves slowly in hydrochloric acid.

11　The interactions between magnetite and other iron-rich oxide minerals—such as ilmenite, hematite, and ulvospinel—have been studied extensively, as the complicated reactions between these minerals and oxygen influence how magnetite preserves records of the Earth's magnetic field.

Uses

12　Magnetite is an important ore of iron.

13　Lodestone, a naturally magnetized form of magnetite, played an important role in the study of magnetism and was used as an early form of magnetic compass.

14　Magnetite typically carries the dominant magnetic signature in rocks, and so it has been a critical tool in paleomagnetism, a science important in discovering and understanding platetectonics. Changes in the oxygen content of the Earth's atmosphere can be inferred by studying sedimentary rocks containing magnetite

15　Igneous rocks commonly contain grains of two solid solutions: one between magnetite and ulvospinel, the other between ilmenite and hematite. A range of oxidizing conditions are found in magmas, and compositions of the mineral pairs are used to calculate how oxidizing the magma was and the possible evolution of the magma by fractional crystallization.

三、分析与讲解

第二课中我们学过，炼铁主要需要的原材料有铁矿石、焦炭、烧结料和石灰石等。烧结料是铁矿石加工后的一种形式，所以主要的原材料其实只有铁矿石、焦炭和石灰石。

铁矿石是含有铁单质或铁化合物的能够经济利用的矿物集合体，是钢铁生产企业的重要原材料。铁矿石的种类很多，用于炼铁的主要有磁铁矿（Fe_3O_4）、赤铁矿（Fe_2O_3）、菱铁矿（$FeCO_3$）、褐铁矿（$Fe_2O_3 \cdot nH_2O$）、钛铁矿（$FeTiO_3$）等。

磁铁矿是铁的氧化物之一，具有亚铁磁性。天然磁化的磁铁矿块，称为磁石。长期以来，关于磁石产生磁性的原因存在三种假说：一些学者认为地球表面的某些东西正在吸引磁石，另一些学者相信磁性来自其他天体，还有一些人认为磁石自身具有磁性。1600 年，医生威廉·吉尔伯特（William Gilbert, 1544—1603）提出整个地球本身就是一块磁铁，这个假设最终通过模型实验得到证实。

我国古代就已经知道如何利用磁力。为了辨明方向，我们的祖先发明了一种罗盘。这种罗盘的指针是一个由磁铁矿制成的指向南方的勺子，勺子放在一个写有文字的盘子上。因为勺子一直指向南方，所以这个物件被称为"司南"。后来经慢慢改进，成为指南针。在我国，指南针被应用于祭祀、礼仪、军事、占卜与看风水等领域。

这篇课文是关于磁铁矿的，重点介绍了磁铁矿的化学成分、特性、用途、储存分布等情况，涉及一些专业术语，翻译时需要细心求证。

第1段：主要是一些专业术语，翻译时不仅需要意思准确，还要表达规范。比如 fracture 这个词可以译为"断裂"，也可以译为"裂缝"，还可以译为"断口"，但在矿物学中应该译成"断口"。再比如 cleavage，可以译为"解理""劈理""卵裂""分裂"等词语，但在矿物学中一般译成"解理"。

在我国，科技术语的规范工作由全国科学技术名词审定委员会（原称"全国自然科学名词审定委员会"）负责。该委员会于1985年经国务院批准成立，是经国务院授权、代表国家审定和公布科技名词的权威性机构。国务院于1987年8月12日明确批示，经全国自然科学名词审定委员会审定公布的名词具有权威性和约束力，全国各科研、教学、生产经营以及新闻出版等单位应遵照使用。1990年，国家科委、中国科学院、国家教委、新闻出版署联合印发通知，提出三项明确要求：①各新闻单位要通过各种传播媒介宣传名词统一的重要意义，并带头使用已公布的名词。②各编辑出版单位今后出版的有关书刊、文献、资料，要求使用公布的名词。特别是各种工具书，应把是否使用已公布的规范名词作为衡量该书质量的标准之一。③凡已公布的各学科名词，今后编写出版的各类教材都应遵照使用。

截至2021年10月，全国科学技术名词审定委员会组建了科学技术各领域名词审定委员会133个，审定公布天文学、物理学、生物化学、电子学、农学、医学、语言学、教育学等140余种规范名词，内容覆盖基础科学、工程与技术科学、农业科学、医学、人文社会科学、军事科学等各个领域。

全国科学技术名词审定委员会建立了权威的术语知识服务平台——术语在线（termonline.cn），通过网络向社会提供公益性科学技术名词查询服务。"术语在线"已成为全球中文术语数据最全、数据质量最高、功能最完备的专业术语知识服务平台，是科技翻译工作者的重要帮手。科技翻译中，凡是不确定术语如何翻译的，建议都先查一下"术语在线"。

全国科学技术名词审定委员会冶金学名词审定委员会于1992年3月成立，第一届委员会由魏寿昆院士任主任委员。从1992年至1999年，完成了《冶金学名词（第一版）》的审定工作，共4917条名词，于1999年正式公布。2007年，第二届冶金学名词审定委员会成立，翁宇庆院士任主任委员。2019年，完成了第二版《冶金学名词》的审定工作，共计

8647 条名词。

第 2 段：(1)科技汉语中也常用化学式，所以像 Fe_3O_4、FeO、Fe_2O_3 这样的化学式都无须翻译。同样，因为 iron(Ⅱ, Ⅲ) oxide 就是 IUPAC 给磁铁矿的正式名称，所以 iron(Ⅱ, Ⅲ) oxide 也无须翻译。如果觉得有必要，可以用括号带出汉语名称"四氧化三铁"。(2)was used as an early form of magnetic compass 字面意思是"被用作磁罗盘的一种早期形式"，也就是"被用在早期的磁罗盘上"。

IUPAC 是 International Union of Pure and Applied Chemistry 的简称，译成汉语为"国际纯粹与应用化学联合会"，又译"国际理论(化学)与应用化学联合会"。这是一个致力于促进化学发展的非政府组织，也是各国化学会的联合组织，以公认的化学命名权威而著称。命名及符号分支委员会会修改 IUPAC 命名法，以力求提供化合物命名的准确规则。这里的 IUPAC 可以①翻译成汉语，②直接照抄，③先照抄，然后用括号带出汉语全称。如果为了照顾读者，当然最后一种方式最可取。

第 3 段：(1)第 1 句意思简单，但表达不易，不妨按照汉语习惯，译为"磁铁矿含铁，富有价值(或：极富价值)"。(2) Plate tectonics(板块构造学)是在大陆漂移学说(Continental Drift)和海底扩张学说(Sea-floor Spreading Hypothesis)的基础上提出的。根据板块构造学，地球表面覆盖着内部相对稳定的板块(plates)，这些板块以每年一定的速度在移动。(3) Sedimentary rocks 意为"沉积岩"，三类岩石中的一种，指由成层堆积于陆地或海洋中的碎屑、胶体和有机物等疏松沉积物团结而成的岩石。另外两种为岩浆岩(也叫火成岩，igneous rocks)和变质岩(metamorphic rocks)，前者指岩浆喷出地表或侵入地壳冷却凝固所形成的岩石，后者指地壳中的原岩(包括岩浆岩、沉积岩和已经生成的变质岩)由于地壳运动、岩浆活动等所造成的物理和化学条件的变化在固体状态下发生了岩石结构、构造甚至矿物成分改变而形成的一种新的岩石。

科技翻译不仅要求译者有较高的双语水平，还要求译者掌握充分的科技知识。冶金生产涉及采矿、选矿、机械、电子、自动控制、物理、化学、物流等多个学科，冶金科技译者也必须拥有这些学科的知识。

第 6 段：A vast deposit of magnetite-bearing sand dunes 中 dunes 是复数，所以应译为"储量巨大的多个含磁铁矿的沙丘"。

第 7 段：New York 当然不是纽约市，而是纽约州。

同一个专有名词，可能会指不同的地方，翻译时需要慎重。比如美国有两个 Washington，一个是 Washington DC，另一个是 State of Washington。前者即美国首都"华盛顿哥伦比亚特区"，靠近弗吉尼亚州(Virginia)和马里兰州(Maryland)，位于美国的东北部、中大西洋地区；后者是位于美国西北部的一个州，州首府为奥林匹亚(Olympia)，最大城市为西雅图(Seattle)。华盛顿州北接加拿大的不列颠哥伦比亚省(British Columbia)，南接俄勒冈州(Oregon)，东接爱达荷州(Idaho)，西邻太平洋。还比如，英国有一个

Cambridge，就是著名的剑桥大学（University of Cambridge）所在地，一般译为"剑桥"，也就是徐志摩《再别康桥》中的"康桥"；美国也有 Cambridge，是美国马萨诸塞州（Massachusetts）波士顿市（Boston）紧邻的一个市，与波士顿市区隔查尔斯河相对，属于波士顿都市区，是哈佛大学（Harvard University）和麻省理工学院（Massachusetts Institute of Technology）的所在地，一般译为"坎布里奇"。

第 10 段：Curie temperature（居里温度）是 19 世纪末法国著名物理学家皮埃尔·居里（Pierre Curie）在自己的实验室里发现的磁石的一个物理特性，即当磁石加热到一定温度时，原来的磁性就会消失。后来，人们把这个温度叫"居里温度"或"居里点"（Curie point）。铁磁物质被磁化后具有很强的磁性，但随着温度的升高，金属点阵热运动的加剧会影响铁磁物质的有序排列。当温度达到足以破坏磁畴磁矩的整齐排列时，磁畴被瓦解，平均磁矩变为零，铁磁物质的磁性消失，变为顺磁物质。与磁畴相联系的一系列铁磁性质全部消失，与铁磁性消失时所对应的温度即为"居里温度"。

第 15 段：solid solutions（固溶体）指溶质原子溶入溶剂晶格中而仍保持溶剂类型的合金相，通常以一种化学物质为基体溶有其他物质的原子或分子所组成的晶体，在合金和硅酸盐系统中较多见，在多原子物质中亦存在。Solid solutions 后面的 one between magnetite and ulvospinel, the other between ilmenite and hematite，one 和 the other 当然指固溶体，一个是磁铁矿和钛尖晶石间的固溶体，另一个是钛铁矿和赤铁矿间的固溶体，可以翻译为"火成岩通常包含两种固溶体的颗粒：一种是磁铁矿和钛铁尖晶石之间的固溶体，另一种是钛铁矿和赤铁矿之间的固溶体"。

四、参考译文

磁 铁 矿

1 类别：矿物

化学式：Fe_3O_4

颜色：黑色，灰色

晶体习性：八面体，细粒状到块状

晶系：等轴

解理：无

断口：不平滑

莫氏硬度：5.5-6.5

光泽：金属光泽

折射率：不透明

条痕：黑

比重：5.17-5.18

2 磁铁矿为铁磁性矿物，化学式为 Fe_3O_4，通用化学名为四氧化三铁，显示磁铁矿不仅含二价铁 FeO（方铁体），也含三价铁 Fe_2O_3（赤铁体）。磁铁矿是几种铁氧化物中的一种，正式（IUPAC）名称为 iron（Ⅱ，Ⅲ）oxide（四氧化三铁）。属尖晶石类矿物，晶体黑色，不透明，呈立方体和八面体。磁铁矿在地球自然存在的矿物中磁性最强，在某些地区以天然磁化的石头"天然磁石"形式存在，被用在早期的磁罗盘上，可缓慢溶解于盐酸。

3 磁铁矿含铁，富有价值。此外，因为磁铁矿带有岩石主要的磁场特征，能将最后一次硬化时的磁象锁定，所以在研究板块构造时发挥着关键的作用。可以通过研究含磁铁矿的沉积岩，推断地球大气层氧气含量的变化。另外，通过研究磁铁矿和其他氧化物矿之间的相互作用，可以确定地质史上岩浆的氧化环境和演化。

4 磁铁矿的细小颗粒几乎存在于所有的火成岩和变质岩中。在沉积岩包括条带状含铁建造中，也可以发现磁铁矿。在很多火成岩里，可以发现从岩浆中沉淀下来的富含磁铁矿和钛铁矿的颗粒。

存象（埋藏）

5 很多沉积岩中都可以找到磁铁矿，条带状含铁建造中层找到了大型储量的矿床。此外，几乎所有的火成岩和变质岩中都有这种矿石（尤其是矿石细粒），橄榄岩和纯橄榄岩的蛇纹石化也可以生成磁铁矿。

6 有时，大量的磁铁矿也会出现在海边的沙滩里，河流的侵蚀作用将磁铁矿带到海边，海浪和洋流将矿砂聚集起来。这样的磁铁矿砂（也称为"铁砂"或"黑砂"）在很多地方都有，包括加州的海滩和新西兰的西海岸。2005年6月，一个勘探公司（坎德罗资源）在秘鲁发现了储量巨大的多个含磁铁矿沙丘，其中最高的沙丘高于沙漠地表2000多米。这些沙丘占地总计250平方公里，磁铁矿含量为10%。

7 已发现的大型磁铁矿分布在瑞典的基律纳和西澳大利亚的皮尔巴拉。此外，挪威、德国、意大利、瑞士、南非、印度、墨西哥等国也有磁铁矿。美国的磁铁矿主要发现在纽约州（阿迪朗达克地区）、新泽西、宾夕法尼亚、北卡罗来纳、弗吉尼亚、新墨西哥、科罗拉多、犹他州和俄勒冈等地。

生物体内的磁铁矿

8 人们发现，某些细菌（如趋磁磁螺菌）以及蜜蜂、白蚁、某些鸟及人类的大脑都有磁铁矿晶体。这些晶体被认为与感知地球磁场磁极和倾角能力的磁感有关，能帮助导航。此外，石鳖牙齿的齿舌是由磁铁类物质构成的，这让石鳖在动物中独树一帜。这意味着，石鳖的舌头非常耐磨，能从岩石上将食物刮擦下来。生物磁学研究开始于20世纪60年代加州理工学院古生态学家海因茨·洛温斯坦的发现。

实验室制备

9 可以使用莫萨特法(the Massart method)在实验室里制备铁磁流体磁铁矿。使用该法时，将二价氯化铁和三价氯化铁在氢氧化钠溶液中搅拌。

特性

10 磁铁矿在所有已知天然矿物中磁性最强，其居里温度是580℃。在化学性质方面，磁铁矿可缓慢溶解于盐酸。

11 磁铁矿和其他富含铁的氧化物矿——钛铁矿、赤铁矿和钛尖晶石——之间的相互作用得到了广泛的研究，因为这些矿物和氧之间的复杂反应影响了磁铁矿对地球磁场信息的记录。

用途

12 磁铁矿是一种重要的铁矿石。磁石这种天然磁化的磁铁矿在磁学研究中发挥了重要的作用，早期被用在磁罗盘上。

13 磁铁矿通常携带有岩石主要磁特征的信息，因此在发现和理解板块构造的古磁学研究中发挥了关键性的作用。

14 可以通过研究含磁铁矿的沉积岩了解地球大气层中氧气含量的变化。

15 火成岩通常包含两种固溶体的颗粒：一种是磁铁矿和钛铁尖晶石之间的固溶体，另一种是钛铁矿和赤铁矿之间的固溶体。岩浆中存在一系列氧化条件，矿物对的组成用于计算岩浆的氧化程度及其可能的演化。

五、专题讨论

科技术语的翻译

专业的科技信息要靠科技术语才能承载，因此科技术语的广泛使用是科技英语的重要特点之一。科技英语的专业性越强，科技术语的使用就越频繁。自然地，要译好科技英语，首先需要译好科技术语。

(一) 科技术语

科技术语是通过语音或文字来定义科学技术概念的约定性语言符号，是描述某一个学科系统知识的关键词，有严格规定的意义。科技术语也就是我们在第二课"专题讨论"中所说的专业词汇及一些半专业词汇。科技术语约占科技英语文本的5%至10%，但它们却构成了科技英语与其他文体的根本区别。

科技术语含义精确，专业性强，词义固定，一般以三种形式存在：

单词式，如 magnetite, ilmentite, flux, coolant, alloy, refractory 等；

复合式，如 ironmaking, taphole, downcomer, lodestone, specific gravity, bustle pipe, pig iron 等；

短语式，如 basic oxygen steelmaking, electric arc furnace, non-recovery coke production, fluidized-bed reactor, vacuum arc remelting 等。

复合式科技术语和短语式科技术语的主要区别在于：复合式大多是由两个单词组成的复合词，其含义一般难以从单个单词的词义直接推出；而短语式则是由多个单词组成的短语，其含义一般可以从组成短语的单词的词义推导出来。

科技英语专业性、逻辑性较强，表达正式、精练和准确。即使是同一个词，在不同学科中，其含义也可能有所不同，所以科技术语的翻译更须注重精确性。

(二) 科学术语翻译的基本原则

科技术语翻译需要遵循如下基本原则：

1. 科学性原则。有些科技术语，人们对其的认知随着时间的改变而改变，其译法也应随着人们认识的加深而改变。比如 stratosphere 这个气象学术语，最先被译为"同温层"，因为限于当时科技发展水平，人们"误认为在'对流层'之上有一层温度'几乎不随高度而变化'，……后来，科学技术发展了，人们已注意到在对流层(tropopause)以上约50公里高空的通常温度'随高度的增加而递增'，原先探测到的只是其底部的温度"①。基于这种新的认识，英语词典对该词的释义作了修改，汉语的译文也随之变为"平流层"。

同样因为科学认识加深而更改了译名的还有 bombesin 一词，这是该词最先被译成"蛙皮素"(a tetradecapeptide first isolated from the skin of certain frogs...)，后来随着科学技术的进展，certain frogs 经分类学鉴定为"铃蟾"，所以正名为"铃蟾肽"。

科学性原则的另一个要求是：科技术语的翻译必须带有科技的味道，比如石油工程中有个名词 doghouse，过去直译为"狗窝"，既不确切也不雅，现在定名为"井场值班房"。还比如，土木工程中有个名词 catwalk，曾被译为"猫道"，现在译为"施工步道"。

2. 单义性原则。科技术语讲究精确，一个术语只能表达一个概念，反过来，一个概念也只能有一个术语与之相对应。如大气科学中的 nowcasting 曾有"现时预报""现场预报""即日预报""短时预报""临近预报"5个译名，现在统一翻译为"临近预报"。还比如地理

---

① 杨枕旦. stratosphere，从"同温层"到"平流层"[J]. 外语教学与研究，2000(6)：461.

学中的 overland flow，曾被翻译为"坡面水流""坡面漫流""陆面水流""地面径流""表面水流"等词，现在统一译为"坡面流"。

不少英语科技术语刚刚引进我国时，都有几种译法，造成了科技交流的困难。也正是为了解决这个问题，国务院才批准成立了全国科学技术名词审定委员会，负责各学科科技名词的统一、规范工作。译者在翻译科技术语时，也要自始至终重视科技术语译名的统一和规范。

3. 简练性原则。科技术语的使用是为了传递科技信息，所以形式简练是术语翻译的自然要求。比如电子学中的 long range and tactical navigation system，全称译成汉语是"远程战术导弹系统"，现按其缩写词 LORTAN 音译为"罗坦系统"。

正是为了简练，科技英语中不少缩略词翻译时常常直接照抄，比如 BF（blast furnace，高炉）、EAF（electric arc furnace，电弧炉）、DRI（direct reduced iron，直接还原铁）、CFB（circulating fluid bed for the reduction of iron ore，用于铁矿还原的循环流化床）等缩略词一般不翻译。

4. 习惯性原则。有些科技术语，虽然最初的译法并不完全妥当，但由于这些译名沿用已久，如果重译可能会有一定的负面影响，因此继续使用习惯译名。比如英文 robot 被译为"机器人"是不准确的，因为 robot 本质不是"人"而是"机器"，称其为"拟人机"或"智能机"更确切。但是考虑到社会上已普遍接受"机器人"这个称呼，已约定俗成，就不再改译，以免引起混乱。

不过也有人认为，科技术语的翻译还是应该向科学性靠拢，这是基本原则。遇到类似的约定俗成的情况时，可在定名时适当变通，使旧术语逐步向新术语过渡。比如计算机术语 menu 已约定俗成地译为"菜单"，但显然计算机中没有"菜"，从其功能角度来看，译为"选单"可能更科学。考虑到"菜单"这一译法已成习惯，公布该术语规范译法时，可以将"选单"作为正式译名，而将"菜单"作为俗称（不推荐使用）。等到社会认可并接受"选单"这一译法后，再逐步废止"菜单"，这是一种积极稳妥的方法。

5. 系统性原则。有时一个学科内几个术语相互联系，翻译时必须遵循系统性原则，考虑这些术语意义的差别，并通过合适的译法将其异同在目的语中充分表达出来。比如，过去 highway、expressway、freeway 都被译为"高速公路"，现在则分别译为"公路""快速路"和"高速公路"。

再比如，粉末冶金领域，有 binder、dopant、lubricant、plasticizer 这些词汇，它们都属于粉末冶金中的添加剂。虽然这些词汇拼写不同，但按照系统性原则，它们被统一翻译成"……剂"（黏结剂、掺杂剂、润滑剂、增塑剂）。同样，acicular、angular、dendritic、fibrous、flaky、granular、irregular、nodular 都是描写粉末颗粒形状的术语，按照系统性原则，将它们分别译为"针状""角状""树枝状""纤维状""片状""粒状""不规则状""瘤状"。

6. 协调性原则。有些术语在多个学科中都会用到，如 probability 这一概念，在数学中被定名为"概率"。而在物理学及其他学科中,过去多翻译为"几率"或"或然率"。其实"概率""几率""或然率"表达的都是一个意思,在这种情况下,就要按主学科的命名方式来统一命名,即统一使用"概率"。这种按主学科来翻译术语的原则就是"协调性原则"。

7. **无歧义原则**。科技术语翻译应避免让译语读者产生歧义的联想。比如 AIDS 这一术语,全称是 acquired immune deficiency syndrome(获得性免疫缺陷综合征),曾被译为"爱滋病",但这种译法容易让人误解为这是一种因爱而滋生的疾病。为了避免这一误解,后来改译为"艾滋病",更准确且无歧义。

总之,英语科技术语汉译,应充分利用汉语词汇的构成规律,体现出汉字和汉语表意性的特点,做到简明扼要、意蕴深远,从形、声、意多方面反映出不同事物之间的差异,表达出事物每个术语的特性和属性。这方面优秀的译例如物理学的"衍射"(diffraction)与"折射"(refraction)。

据说,1933 年 8 月 21 日至 9 月 2 日,在上海中央研究院物理研究所召开了一次名词审查会议。时任物理研究所研究员的杨肇燫对此工作非常积极,每日必到,并深入研究。当时讨论 damping 一词的翻译,有"减幅""阻迟"等说,但总觉得不够妥当。第二天继续开会,杨肇燫一到会即说,他昨夜忽得一"尼"字,有逐步减阻之意。大家交相称赞,于是将 damping 的中译文定为"阻尼"。这一译法至今仍被采用,成为一段佳话。

大气科学中有 mirage 一词,意为"空气光线穿过密度梯度足够大的近地气层时,光线发生显著折射,从而在空中或地平线下出现的奇异幻景"。由于中国早有"海市蜃楼"之说,故将此术语定名为"蜃景"。

(三)科学术语翻译的常用方法

同其他词汇一样,科技英语的翻译方法主要包括直译、意译、音译等几大类。具体到操作层面,又有不同的处理手段:

1. 移植。移植就是按词典里所给的词义将术语各个词素的意义依次译出,这是一种严格的直译方法。移植多用于派生词和复合词,如:inclusion distribution(夹杂物分布)、inner segregation(内部偏析)、kissing point(吻点)、mechanical automation(机械自动化)、microwave(微波)、superconductor(超导体)、magnetohydrodynamics(磁流体力学),等等。这些术语往往由一些基本的科技英语词素组合而成,因而直接按原术语顺序进行翻译即可。

2. 推演。推演指根据术语在原文或原语词典中的意思,推演出汉语的译义。译语不仅包含原术语的字面意义,还必须概括出术语所指事物的基本特征,属于意译。如 space shuttle 一词,如果采用移植将其译成"太空穿梭机",很容易引起误解。其实,这里的 space 指的是 aerospace(航天),shuttle 指往返于太空与地球之间、形状像飞机的交通工具,因此,将 space shuttle 推演译成"航天飞机",不仅直观易懂,也更容易被人们接受。推演使用得当，能得到高质量的译文，但要求译者有较好的专业知识和两种语言的良好素养。

推演时有时也可以从反向着手。比如：undesired signal 字面意思是"不受欢迎的信号"，实际应译为"干扰信号"。类似的例子还有：undesired sound（噪声）、unenclosed construction（敞开式结构）、optional parts（非保准件）、quiet circuit（无噪声电路）、quiet run（无声运转）、shutdown（非工作周期）、restricted motion（非自由运动）。

英语中还有一类术语，其真正意思与字面意思相反，也需要从反向着手进行翻译。比如：cough syrup 字面意思是"咳嗽糖浆"，当然我们不会生产"让人咳嗽的糖浆"，只能是"止咳糖浆"。类似的例子还有：dusk mask（防尘罩）、wear pump（耐磨泵）、wear washer（耐磨垫圈）、wet strength（耐湿强度）、compression test（耐压试验）、shear strength（抗剪强度）、tear test（抗扯试验）、explosive starter（防爆启动器），等等。

3．引申。引申是在不脱离原文的基础上，运用延续与扩展的方法译出原术语，与推演相比又向前进了一步，也是意译的一种。引申可以将具体所指引向抽象泛指，如 head 是"头、头部"的意思，抽象意义为"先导、领先"，head dislocation 则可以引申译为"先导位错、领先位错"。引申也可以将抽象泛指引向具体所指，如 career 抽象泛指"事业、职业、生涯"，当用于指高炉、转炉等时，则引申至具体的"炉龄、高/转炉寿命"。

4．解释。解释法即用目的语给出原术语的意思而不给出对等词，还是属于意译。例如：blood type 可译为"血型"，blood bank 可译成"血库"，但 blood heat 却不能译成"血热"，而采用其他方法也很难译出该术语的准确含义。此时可借用解释法，将其译成"人体血液正常温度"。解释法多用于初次出现且意义比较抽象、含义比较深刻的术语。

5．音译。有些科技术语中在目的语中没有确切的对等词，按照意译又比较费劲，只好借助于音译。例如：hacker（黑客）、nylon（尼龙）、aspirin（阿斯匹林）、radar（雷达）、joule（焦耳）、hertz（赫兹）、ampere（安培）、volt（伏特），等等。

总体说来，中国人对音译的接受度比意译低，有不少术语刚开始是音译，后来慢慢改成了意译，这样的例子如：shock（休克→虚脱）、engine（引擎→发动机）、email（伊妹儿→电子邮件/电邮）、cement（水门汀→水泥）、film（菲林→胶卷）、hormone（荷尔蒙→激素），等等。

有学者通过对《科技标准术语词典》（1—7 卷）中所收录的音译科技术语进行梳理，发现该词典共收录理工农医等各类专业术语 77000 余条，其中采用音译法进行翻译的术语有 716 条，占比约 0.09%。采用音译的术语主要分为以下六类：（1）医药或化学品名称，如 pyridine（吡啶），furan（呋喃）；（2）新发现或新物质名称，如 quark（夸克）、valitin（凡立丁）；（3）计量单位名称，如 bit（比特）、neper（奈培）；（4）新技术、新工艺名称，如 clone（克隆）、Bengough-Stuart process（本高-斯托特工艺）；（5）新设备、新系统名称，如 Derimotor（戴利电动机）、Consol（康索尔）；（6）以科学家姓氏或名字命名的术语，如

Poisson's ratio(泊松比)、Cherenkov effect(契伦柯夫效应)。①

这位学者提出，音译可分为纯音译、谐音译、形音译、音意兼译四类。上述例子中的"Bengough-Stuart process(本高—斯托特工艺)"就属于音意兼译。钢铁冶金学科中音意兼译的例子特别多，如：Abel's reagent(阿贝尔侵蚀剂), Acheson furnace(艾奇逊电炉), Babbitt alloy(巴比合金), Baker dam(贝克撇渣器), Calderon charging machine(卡尔德龙废钢装料机), Cape-Brasst process(凯普—布拉塞特直接炼铁法), 等等。

谐音译一般有着特别的目的，比如 rogor 是一种含毒量低的杀虫剂，可用于果树、蔬菜，谐音译作"乐果"，能缓解人们对该物品毒性的抵触。又如 sofar channel(声发声道), 选"声发"二字谐音译 sofar, 能让人立刻联想到该术语和声音有关，进而记住这一术语的所指"深海声道"。

术语音译还常利用方言谐音，如"的确良"(dacron)就是采用粤语谐音译出的。又如"乌龙球"来自 own goal 的谐音译，因为 own goal 与粤语的"乌龙"发音相近。

至于这位学者提到的形音译，我们认为应归入形译，如：A-drive(A式传动), C-hanger(C形吊具), O-ring(O型密封圈), F-stop(F制光圈), V-step(V字步), X-ray(X射线), ID coil(ID线圈), PH value(PH值), Tab character(Tab字符), PCV valve(PCV阀), K frequency band(K波段), ABS plastic(ABS塑料), DNB ratio(DNB比), I/O controller(I/O控制器), AAC-system(AAC系统), UCS system(UCS色系), RGB signal(RGB信号), 等等。

6. 形译法。所谓形译，就是使用字母或汉字来表达原术语中展现的形状。形译法有以下两种情况：

第一种，保留原术语中的字母不译，如：S-turning(S形弯道), A-bedplate(A形底座), T-bolt(T字螺栓)。

第二种，用汉语中表示相似形象的词来译，如：steel I-beam(工字钢梁), OV ring(环形圈), L-square(直角尺), U-bolt(马蹄螺栓), V-belt(三角带), T-bend(三通接头), twist drill(麻花钻), 等等。

科技英语表达严谨、规范，科技英语翻译不要求华丽辞藻的堆砌，只要求准确、客观地传达原文承载的知识和信息。只有掌握科技术语的译法，才能将科技英语翻译得更准确、更严密和更完整。

◎ 练习三　将下面的句子翻译成汉语。

1. A piece of magnetite has magnetic properties and attracts iron or steel. Iron ores of greatest economic interest contain magnetite and hematite.

---

① 阳琼. 大数据下的科技术语音译[J]. 中国科技翻译 2018(3)：1-3.

2. Strictly speaking, steel is just another type of iron alloy, but it has a much lower carbon content than cast iron and roughly the same (or sometimes slightly more) carbon than wrought iron, and other metals are often added to give it extra properties.

3. According to the mineralogical characteristics of the low-grade fine particle magnetite ore in Inner Mongolia, in order to reduce production costs, high degree iron concentrate was obtained by the process of stage grinding and stage magnetic separation.

4. With the more complex shapes and when less ductile and malleable materials are used, it is necessary to go through a series of stages before the shape can be produced from the raw material.

5. The rusting of iron is only one example of corrosion, which may be described as the destructive chemical attack of a metal by media with which it comes in contact, such as moisture, air and water.

6. The iron, as it were, breathes air as we do, and as it breathes, softening from its merciless hardness, it falls into fruitful and beneficent dust, gathering itself again into the earth from which we feed and the stones with which we build, into the rocks that frame the mountains and the sands that bound the sea.

7. Bats famously orientate at night by echolocation, but this works over only a short range, and little is known about how they navigate over longer distances. Here we show that the homing behaviour of Eptesicus fuscus can be altered by artificially shifting the Earth's magnetic field, indicating that these bats rely on a magnetic compass to return to their home roost.

8. Plate Tectonics refers to the study of the structure of the earth's crust and mantle with reference to the theory that the earth's lithosphere is divided into large rigid blocks (plates) that are floating on semifluid rock and are thus able to interact with each other at their boundaries, and to the associated theories of continental drift and seafloor spreading.

9. A range of oxidizing conditions is found in magmas, and compositions of the mineral pairs are used to calculate how oxidizing the magma was and the possible evolution of the magma by fractional crystallization.

10. Moreover, interactions between magnetite and other oxide minerals have been studied to determine the oxidizing conditions and evolution of magmas over geological history.

第四课　高炉炼铁用焦炭的生产

> 北京诸山多石炭,俗称水火炭,可和水而烧之也,……或炼焦炭,备冶铸之用。
>
> ——明·李诩(1505—1593)《戒庵漫笔》

一、科技术语

aromatic hydrocarbon compounds Organic compounds containing one or more aromatic rings.(芳香烃化合物)

blast furnace A type of metallurgical furnace used for smelting to produce industrial metals, generally pig iron, but also others such as lead or copper. "Blast" refers to the combustion air being "forced" or supplied above atmospheric pressure. (高炉)

bulk density A property of powders, granules, and other "divided" solids, especially used in reference to mineral components (soil, gravel), chemical substances, (pharmaceutical) ingredients, foodstuff, or any other masses of corpuscular or particulate matter (particles).(散装密度;堆密度)

burden Raw materials used in a blast furnace, such as coke, ore and limestone.(高炉炉料;装料)

by-product cokemaking A cokemaking process with byproducts recovered for recycling.(副产品回收炼焦工艺)

carbonization The destructive distillation of coal (as in coke ovens).(干馏;碳化)

charge level A level in the blast furnace or converter above which there should be no charge. (装料线)

coal blend The process of mixing coals after coal has been mined to achieve quality attributes that are desirable for the coal's intended application such as cokemaking.(配煤)

hot metal Pig iron in a liquid state.(铁水)

non-recovery/heat recovery coke production A process for cokemaking in which no heat or heat is recovered.(无热回收/热回收炼焦工艺)

quench A rapid cooling process of burning coke in the materials science. (息焦)

reduction The fact of adding one or more electrons to a substance or of removing oxygen from a substance. (还原)

二、英语原文

Coke Production For Blast Furnace Ironmaking[①]

Hardarshan S. Valia(Scientist, Ispat Inland Inc.)

Introduction

1 A world class blast furnace operation demands the highest quality of raw materials, operation, and operators. Coke is the most important raw material fed into the blast furnace in terms of its effect on blast furnace operation and hot metal quality. A high quality coke should be able to support a smooth descent of the blast furnace burden with as little degradation as possible while providing the lowest amount of impurities, highest thermal energy, highest metal reduction, and optimum permeability for the flow of gaseous and molten products. Introduction of high quality coke to a blast furnace will result in lower coke rate, higher productivity and lower hot metal cost.

Coke Production

2 The cokemaking process involves carbonization of coal to high temperatures (1,100℃) in an oxygen deficient atmosphere in order to concentrate the carbon. The commercial cokemaking process can be broken down into two categories: a) By-product Cokemaking and b) Non-Recovery/Heat Recovery Cokemaking. A brief description of each coking process is presented here.

By-product Cokemaking

3 The majority of coke produced in the United States comes from wet-charge, by-product

---

① 本文来源于 https://accci.org/wp-content/uploads/2021/07/coke-production-for-blast-furnace-ironmaking-07-22-2021.pdf,有删改。

coke oven batteries (Figure 4.1). The entire cokemaking operation is comprised of the following steps: before carbonization, the selected coals from specific mines are blended, pulverized, and oiled for proper bulk density control. The blended coal is charged into a number of slot type ovens wherein each oven shares a common heating flue with the adjacent oven. Coal is carbonized in a reducing atmosphere and the off-gas is collected and sent to the by-product plant where various by-products are recovered. Hence, this process is called by-product cokemaking.

Figure 4.1 "Coke Side" of a By-Product Coke Oven Battery (The oven has just been "pushed" and railroad car is full of incandescent coke that will now be taken to the "quench station")

图 4.1 副产品回收炼焦炉组的"焦侧"(该焦炉刚刚推焦,火车车皮装满了炽热的焦炭,将其送往熄焦站)

图源:https://ar.inspiredpencil.com/pictures-2023/coke-oven-process

4 The coal-to-coke transformation takes place as follows: the heat is transferred from the heated brick walls into the coal charge. From about 375℃ to 475℃, the coal decomposes to form plastic layers near each wall. At about 475℃ to 600℃, there is a marked evolution of tar, and aromatic hydrocarbon compounds, followed by resolidification of the plastic mass into semi-coke. At 600℃ to 1100℃, the coke stabilization phase begins. This is characterized by contraction of coke mass, structural development of coke and final hydrogen evolution. During the plastic stage, the plastic layers move from each wall towards the center of the oven trapping the liberated gas and creating in gas pressure build up which is transferred to the heating wall. Once, the plastic layers have met at the center of the oven, the entire mass has been carbonized. The incandescent coke mass is pushed from the oven and is wet or dry quenched prior to its shipment to the blast furnace.

Non-Recovery/Heat Recovery Coke Production

5 In Non-Recovery coke plants, originally referred to as beehive ovens, the coal is carbonized in large oven chambers (Figure 4.2). The carbonization process takes place from the top by radiant heat transfer and from the bottom by conduction of heat through the sole floor. Primary air for combustion is introduced into the oven chamber through several ports located above the charge level in both pusher and coke side doors of the oven. Partially combusted gases exit the top chamber through "down comer" passages in the oven wall and enter the sole flue, thereby heating the sole of the oven. Combusted gases collect in a common tunnel and exit via a stack which creates a natural draft in the oven. Since the by-products are not recovered, the process is called Non-Recovery cokemaking. In one case, the waste gas exits into a waste heat recovery boiler which converts the excess heat into steam for power generation; hence, the process is called Heat Recovery cokemaking.

Figure 4.2 Heat recovery coke plant

图4.2 热回收焦化厂

图源：清洁型热回收焦炉图册_360百科 https://baike.so.com/gallery/list? ghid = first&pic _idx = 1&eid = 832831&sid = 880826

Coke Properties

6 High quality coke is characterized by a definite set of physical and chemical properties

that can vary within narrow limits. The coke properties can be grouped into following two groups: a) Physical properties and b) Chemical properties.

Physical Properties

7 Measurement of physical properties aid in determining coke behavior both inside and outside the blast furnace. In terms of coke strength, the coke stability and Coke Strength After Reaction with CO_2(CSR) are the most important parameters. The stability measures the ability of coke to withstand breakage at room temperature and reflects coke behavior outside the blast furnace and in the upper part of the blast furnace. CSR measures the potential of the coke to break into smaller size under a high temperature CO/CO_2 environment that exists throughout the lower two-thirds of the blast furnace. A large mean size with narrow size variations helps maintain a stable void fraction in the blast furnace permitting the upward flow of gases and downward of molten iron and slag thus improving blast furnace productivity.

Chemical Properties

8 The most important chemical properties are moisture, fixed carbon, ash, sulfur, phosphorus, and alkalies. Fixed carbon is the fuel portion of the coke; the higher the fixed carbon, the higher the thermal value of coke. The other components such as moisture, ash, sulfur, phosphorus, and alkalies are undesirable as they have adverse effects on energy requirements, blast furnace operation, hot metal quality, and/or refractory lining. Coke quality specifications for one large blast furnace in North America are shown in Table 4.1.

Factors Affecting Coke Quality

9 A good quality coke is generally made from carbonization of good quality coking coals. Coking coals are defined as those coals that on carbonization pass through softening, swelling, and resolidification to coke. One important consideration in selecting a coal blend is that it should not exert a high coke oven wall pressure and should contract sufficiently to allow the coke to be pushed from the oven. The properties of coke and coke oven pushing performance are influenced by following coal quality and battery operating variables: rank of coal, petrographic, chemical and rheologic characteristics of coal, particle size, moisture content, bulk density, weathering of coal, coking temperature and coking rate, soaking time, quenching practice, and coke handling. Coke quality variability is low if all these factors are controlled. Coke producers use widely differing coals and employ many procedures to enhance the quality of the coke and to enhance the coke oven productivity and battery life.

Table 4.1 Coke Quality Specifications

Physical (measured at the blast furnace)	Mean	Range
Average Coke Size (mm)	52	45-60
Plus 4" (% by weight)	1	4 max
Minus 1" (% by weight)	8	11 max
Stability	60	58 min
CSR	65	61 min
Chemical (% by weight)	Mean	Range
Ash	8.0	9.0 max
Moisture	2.5	5.0 max
Sulfur	0.65	0.82 max
Volatile Matter	0.5	1.5 max
Alkali (K_2O+Na_2O)	0.25	0.40 max
Phosphorus	0.02	0.33 max

三、分析与讲解

文章介绍高炉炼铁所需的重要原材料焦炭的生产，内容包括焦炭对于高炉炼铁的重要意义，焦炭的主要生产工艺、性质以及影响焦炭性质的主要因素等。总体而言，内容不难，所述过程较简单，基本理解应该不成问题，但要做到准确流畅地翻译，还需注意以下几个方面：

1. **准确翻译科技术语**。如同第三课中所说，术语翻译事关科技信息的准确传达，以科学性、简练性、系统性等为原则。具体到本课中出现的术语，特别需要关注那些与普通词汇没有差异的表达式，不能将专门术语当成普通词汇来翻译，如 hot metal 不是"热金属"而是"铁水"，burden 不是"负担"而是"炉料"，charge 不是"指控"或"要价"而是"装料"，coke over batteries 不是"焦炉电池"而是"焦炉组"，等等。这些术语在冶金科学中意义固定，译者不仅应认识这些术语，还应了解术语的内涵，比如 coke rate（焦比）是冶炼一炉铁水所用的焦炭和产出铁水的比值。

2. **灵活运用翻译技巧**。汉、英双语差异太大，翻译时，译者不应被原文使用的词汇和句法结构束缚，而应以准确表达原文内涵为第一要义。因此，可以根据需要使用包括增词、减词、词性转换、次序调整、正说反译和反说正译法等翻译技巧。请看第1段的最后一句：

Introduction of high quality coke to a blast furnace will result in <u>lower</u> coke rate, <u>higher</u> productivity and <u>lower</u> hot metal cost.

在高炉中使用高质量的焦炭可以<u>降低</u>焦比，<u>提高生产率</u>，<u>减少铁水生产成本</u>。

原文 result in 后有三个并列宾语，直译便是："……将带来更低的焦比、更高的生产率和更低的铁水成本。"虽然意思是对的，但不太符合汉语的表达习惯，语言也缺乏力量和气势。如将 lower、higher 和 lower 三个形容词比较级译成动词，可以明显改善句子的节奏，表达也更有力度。

再请看第7段的最后一句：

<u>A large mean size</u> with <u>narrow size variations</u> helps maintain a stable void fraction in the blast furnace…

平均粒度较大、粒度均匀的焦炭有助于高炉保持一个稳定的空隙组分……

原句主语是一个带后置介词短语的名词短语，如直译将过于烦琐。此处，译者将 a large mean size, with narrow size variations 分别视为 mean size is large, size variations are narrow 两个命题，翻译时，将命题内容以肯定方式译出，并补充主语"焦炭"，这样用两个命题组成的定语修饰主语，不但语义明确，也更加符合科技文本语言简洁明了的要求。

冶金科技翻译中，需要灵活使用翻译技巧的例子大量存在。译者不应只关注形式上的对等，还应致力于译出意义对等且表达形式更符合汉语习惯的句子。

3. 合理转换长短句。英语习惯使用各种从句以达到表达精确的目的，汉语则习惯使用小句，或者叫短句。翻译时，应根据原文各成分间的逻辑关系将英语长句分成不同的意群，然后以中文短句的形式表达出来，实现语义对等且语言形式不拗口的目的。比如第3段，叙述了焦炭在焦炉中受热变化的过程，其中有很多长句。汉译时，应以主要动词为中心，将长句转化成汉语的小句。再看第4段中的一句：

During the plastic stage, the plastic layers <u>move</u> from each wall towards the center of the oven <u>trapping</u> the liberated gas and <u>creating</u> in gas pressure build up which is <u>transferred</u> to the heating wall.

在胶质层阶段，胶质层从墙边向炉孔中心<u>移动</u>，<u>捕获</u>释放的煤气，<u>造成</u>煤气压力增加，并将压力<u>转给燃烧室</u>。

原句描述的是受热时胶质层的一系列变化，过程性非常强，每一步的变化又带来新的变化，分别用动词 move，trapping，creating 和 is transferred 来表达。从 trapping 开始，每个动词描述了胶质层的一个阶段性变化，类似于汉语的流水句。翻译时，应分清动作顺序，以并列形式表达时间上的延续："……胶质层从墙边向炉孔中心<u>移动</u>，<u>捕获</u>释放的煤气，<u>造成</u>煤气压力增加，并将压力<u>转给</u>燃烧室。"类似情况在第 5 段描述另外一种炼焦工艺时，也很多。

4. **慎重处理出现的介词**。英汉翻译时，介词有时需要翻译，有时不需要翻译，需斟酌决定，比如第 5 段中间一句：

Primary air <u>for</u> combustion is introduced <u>into</u> the oven chamber <u>through</u> several ports located <u>above</u> the charge level <u>in</u> both pusher and coke side doors <u>of</u> the oven.

一次助燃空气<u>通过</u>炉料<u>上方</u>机侧和焦侧<u>炉门上</u>的进气孔<u>进入</u>炉室。

英语中的介词通常具有重要的语法功能，表明名词概念之间的相互关系，如目的（for）、趋向（into）、方式（through）、方位（above、in）、所属（of）等。其中有些可以隐含在汉语表达中，无须译出，有些因为同时具有很强的信息性，必须作显化处理。所以，翻译时须灵活选择翻译策略。比如介词 for 和 of，在"一次助燃空气""焦侧炉门"中已经隐含，无须译出；而其他几个介词的意思因为涉及方位、趋向、方式等具体信息，不译则会造成理解偏差或者理解不足，属于有必要显化的信息，需明确翻译出来。

下面将逐段分析：

第 1 段：(1)原文中 the highest quality 同时描述 raw materials，operation 和 operators，翻译时需要使用不同的形容词："高质量的原材料""良好的操作""熟练的操作工"。(2)第 2 句是个结构复杂的简单句，翻译时最好将状语 in terms of its effect on blast furnace operation and hot metal quality 提前。(3)第 3 句结构同样复杂，最好将优质焦炭的几个重要作用翻译成并列动词短语："保证炉料的顺利下降""尽量减少炉料的劣化""降低杂质的含量""提供充足的热能""最大程度地还原金属""为高炉气态和液态产品的流动提供良好的透气性"。

第 2 段：第 1 句是对 The cokemaking process 下定义，翻译时最好不要断句，而是按照原句语序直接翻译成"炼焦就是……的过程"，这样才显得完整、严密。

第 3 段：(1)第 1 句：wet-charge 指"装入焦炉的煤炭含有水分"，专业表述是"湿煤装料"。(2)后面几句中术语的翻译同样需要准确：配煤（blended）、粉碎（pulverized）、加油（oiled）、堆密度控制（bulk density control）、室式焦炉（slot type oven）、火道（heating flue），建议查阅专门的冶金词典，如徐树德、赵予生主编的《英汉钢铁冶金技术详解词典》(机械工业出版社,2015年1月第1版)。

第4段:(1)翻译 the heat is transferred from the heated brick walls into the coal charge 时,应补充 the heat 来自何方的信息,即"燃烧室燃烧产生的热量",否则会给读者突兀的感觉。一般来说,焦炉组是碳化室和燃烧室依次相间的布局,即一个碳化室紧邻一个燃烧室,然后又是碳化室……燃烧室产生的热量通过砖墙传到隔壁的碳化室,碳化煤炭,使之成为焦炭。(2)翻译 This is characterized by… final hydrogen evolution 时,应使用词性转换的翻译技巧,将 characterized 译成名词,而将 contraction、development、evolution 三个名词译成动词。(3)碳化室内,the plastic layer 是从靠近燃烧室的四边开始形成,然后向中间移动,所以才有 the plastic layers move from each wall towards the center of the oven。

第5段:(1)heat transfer 和 conduction of heat 有区别,分别是"热传递"和"热传导"。Heat transfer 发生在焦炉上部有空隙的区域,不接触传热,而 conduction of heat 发生在焦炉底部,热量来自被加热的炉底,炉料与炉底直接接触。(2)最后一句 In one case,说的是"在一种情形下",也就是"也有可能"。

第6段:第1句中 a definite set of 不能译为"一系列确定的",因为焦炭是大宗工业品,不可能所有的焦炭——即使是高质量焦炭——都有一样的确定的物化性质。翻译时不妨语气柔和一些,译为"一系列较为固定的"。

第7段:(1)因为第2句提到了 coke stability 和 CSR,因此第3、4句中的 The stability 和 CSR 可以分别译为"前者""后者",不仅简洁,而且前后连贯。(2)第4句比较长,比较好的译法是将 a high temperature CO/CO_2 翻译成外位成分,然后用"这样的环境"作为接下来一句的主语。关于外位句,请参看第八课的"专题讨论"。

第8段:(1)翻译第1句时,不能简单地说"焦炭最重要的化学性质是水分、固定碳、灰分、硫和碱金属",而要增加"含有……的量",如此才能体现每一种成分的化学意义。

表4.1:(1)Average Coke Size 中 Average 可以理解为"平均",也可以理解为"普通",也就是绝大多数焦炭粒度为52,但因为已经有一项指标是 Mean(平均值),所以此处译为"普通"较好。(2)Plus 4"和 Minus 1"不好理解,其实就是比 Average Coke Size 大4mm和小1mm的意思。

第9段:(1)第2句给炼焦煤下定义,用了2个动名词(softening、swelling)和1个名词(solidification),描述炼焦过程中焦煤的变化。虽然英语用的是逗号,但译成汉语应改成顿号。英语中没有顿号,所以句内停顿一律用逗号,但汉语标点比较细致,比逗号更短的停顿还有顿号,一般用于词语之间的停顿。同样需要注意标点的是第4句的翻译:battery operating variables 后面跟随的是一系列影响焦炭性质和推焦性能的因素,英语原文在这些因素间用的都是逗号。但考虑到 petrographic、chemical and rheologic characteristics of coal、coking temperature and coking rate 在语义上相对独立,所以翻译时在"岩相学""化学""流变学"以及"炼焦温度""炼焦速率"等中间用顿号,而在语义相对独立的词组间用分号。(2)第4句中的coke handling,指的是"焦炭的装卸"。

四、参考译文

高炉炼铁用焦炭的生产

哈达山·辛格·威利亚

(伊斯帕特内陆股份有限公司旗下科学家)

简介

1 世界级的高炉生产需要高质量的原材料、良好的操作和熟练的操作工。就其对高炉运转和铁水质量的影响而言,焦炭是最重要的高炉生产原料(在装炉物料中最具重要性/最为关键)。高质量的焦炭能够保证高炉炉料的顺利下降,尽量减少炉料的劣化,降低杂质的含量,提供充足的热能,最大程度地还原金属,以及为高炉气态和液态产品的流动提供良好的透气性。在高炉中使用高质量的焦炭可以降低焦比,提高生产率,减少铁水生产成本。

焦炭的生产

2 炼焦就是将煤放在缺氧氛围中(隔绝空气条件下)加热到高温(1100℃)碳化(干馏)以浓缩炭的过程。工业化炼焦分两种工艺:副产回收炼焦和无(副产)回收/热回收式炼焦。这里简要介绍一下这两种工艺。

副产品回收炼焦工艺

3 美国生产的大部分焦炭都来自于湿煤装料副产回收的炼焦炉组(如图4.1所示)。整个炼焦过程包括如下步骤:碳化前,从特定煤矿精选煤炭进行配煤、粉碎和加油,以达到合适的堆密度。配好的煤装进一系列室式焦炉(炉孔)里,每个炉子与相邻的炉子共享(加热)火道。煤炭在还原气氛中碳化,排出的气体被收集起来送到收工厂回收各种副产品。因此,这种工艺被称为副产回收炼焦。

4 煤炭至焦炭转换过程如下:热砖墙将燃烧室燃烧产生的热量传到煤炭炉料。375℃—475℃间,靠近热砖墙的煤分解为胶质层。475℃—600℃时,焦油和芳香烃化合物明显生成,接着胶质层再凝固为半焦。600℃—1100℃时,焦炭开始成形。这个阶段的特征是,焦块收缩,焦炭结构形成,氢气最后逃逸。在胶质层阶段,胶质层从墙边向炉孔中心移动,捕获释放的煤气,造成煤气压力增加,并将压力转给燃烧室。一旦胶质层在炉孔中心相遇，整个煤块已经碳化。炽热的焦块被推出焦炉，在送入高炉前，经过干法熄焦或湿法熄焦。

无(副产)回收/热回收式炼焦工艺

5　无(副产)回收焦炉，原被称为"蜂巢炼焦炉"。在无回收焦炉厂里，煤炭在大型焦炉室里碳化(如图4.2所示)。从上部的辐射热传递和下部炉底的热传导引起煤的碳化。一次助燃空气通过炉料料线上方机侧和焦侧炉门上的进气孔进入炉室。部分燃烧的气体通过安装在炉墙上的"下降管"从最上层炉室逸出，进入炉底烟道，加热炉底。燃烧的气体收集在一个共同的管道中，通过烟囱排出，在炉孔里产生自然风流。因为没有副产品回收，所以该工艺称为无回收炼焦。有的情况下，废气被送进废气余热回收锅炉，生产蒸汽，用于发电，所以该工艺也被称为"热回收炼焦工艺"。

焦炭的性质

6　高质量焦炭具备一系列较为固定的物理和化学性质，这些性质只在很小的范围内波动。焦炭的性质可以归为两类：物理性质和化学性质。

物理性质

7　焦炭物理性质的测定有助于确定焦炭在高炉内外的表现。就焦炭强度而言，焦炭的稳定性和焦炭与二氧化碳反应后的强度是最重要的参数/指标。前者测量的是焦炭在室温下抵抗破碎的能力，反映的是焦炭在高炉外和高炉上部的表现。后者测量的是焦炭在高温且充满CO/CO_2的环境下破碎的可能性，这样的环境出现在高炉下面三分之二的区域。平均粒度较大、粒度均匀的焦炭有助于高炉保持一个稳定的空隙率，便于炉气向上流动及铁水和渣的向下流动，因此能提高高炉生产率。

化学性质

8　焦炭最重要的化学性质包括含有水分、固定碳、灰分、硫、磷和碱金属的量。固定碳是焦炭中燃料所占的比例。固定碳含量越高，焦炭的热值就越高。其他成分如水分、灰分、硫、磷和碱性物质都是人们想避免的，因为它们对满足能量要求、高炉运转、铁水质量和/或耐火炉衬有负面影响。表4.1显示的是北美一家大型高炉对焦炭质量的要求。

表4.1　北美一家大型高炉对焦炭质量的要求

物理性质(高炉内测得)	平均值	区间	化学性质(重量百分比)	平均值	区间
普通焦炭粒度（mm）	52	45—60	灰分	8.0	≤9.0

续表

物理性质(高炉内测得)	平均值	区间	化学性质（重量百分比）	平均值	区间
超过普通焦炭 4mm（重量百分比）	1	≦4	水分	2.5	≦5.0
低于普通焦炭 1mm（重量百分比）	8	≦11	硫	0.65	≦0.82
稳定性	60	>58	挥发性物质	0.5	≦1.5
反应后强度	65	>61	碱性物质（K_2O 和 Na_2O）	0.25	≦0.40
			磷的含量	0.02	≦0.33

影响焦炭质量的因素

9　高质量的焦炭通常来自高质量炼焦煤的碳化。炼焦煤指的是那些碳化时经历软化、膨胀、再固化形成焦炭的煤。选择煤种配煤时，一个重要的考虑因素是该煤种不会对焦炉炉壁产生巨大的压力，而且能够收缩得足够多，以便焦炭顺利推出焦炉。焦炭性质和推焦性能受下述焦炭质量和焦炉操作变量的影响：煤的等级，煤的岩相学、化学和流变学特征，煤的颗粒大小、水分含量、堆密度和风化情况，炼焦温度、炼焦速率、保温时间，熄焦工艺和焦炭的装卸。如果这些因素都得到了控制，那么焦炭质量的变化就小。焦炭生产商使用差异很大的煤炭，采用不同的工艺，提高焦炭的质量，提高焦炉生产率和焦炉炉龄。

五、专题讨论

科技翻译中网络资源的使用

我们早已进入了网络时代，网络上大量免费或收费的资源是我们生活、工作的重要帮手。虽然网络资源有资源分散、查找起来费时费力、可靠性不高等诸多问题，但其优势也是明显的：资源种类繁多，不仅有海量文字资源，还有大量的音频、视频资源；资源时效性强，更新快；可以随时登录网络查找利用；便于计算机处理，方便快捷。正因如此，从事科技翻译的人士理应掌握利用网络资源提高科技翻译质量的方法。

对于科技翻译工作者来说，可资利用的网络资源大致可以分为如下几类：

1. 在线词典。在线词典分多种类型，有单语、双语甚至多语，有通用型、专业型，有文本型、多媒体型，涵盖了几乎所有的学科和专业。由于这些在线词典大多是基于已经出版的纸质词典，所以质量普遍较高，可靠性较强。可以说，有了网络在线词典（包括一些词典类 App），传统的纸质词典几乎无用武之地了。这里列举一些常用的在线词典①：

https：//dictionary.cambridge.org/（Cambridge Dictionaries Online：英英词典，兼具翻译功能）

https：//www.ldoceonline.com/（Longman Dictionary of Contemporary English：英英词典）

https：//www.dictionary.com/（Dictionary.com：英英词典）

https：//www.oxforddictionaries.com/（Oxford Dictionaries：英英词典）

https：//www.macmillandictionary.com/（Macmillan Dictionary：英英词典）

https：//www.thefreedictionary.com/（The Free Dictionary：英英词典、百科查询）

https：//www.wiktionary.org/（Wikitionary：英英词典）

https：//www.iciba.com/（爱词霸：兼具查词和翻译功能，既有单语词典，也有双语词典，还可以提取图片文字并翻译及取词划译等功能。）

https：//dict.cn/（海词：英汉、汉英查词）

https：//dict.youdao.com/（有道：英英释义、英汉释义）

http：//www.dict.li/（里氏词典在线：英汉、汉英查询）

https：//www.dreye.com.cn/index_gb.html（译典通：英汉查询）

https：//www.zdic.net/（汉典：汉语字典、汉语词典、成语词典）

http：//www.hydcd.com/（汉辞网：汉语字典、汉语词典、成语词典、英汉词典）

https：//www.ozdic.com/（OzDictionary：英语搭配词典，可以查询自己使用的英语搭配是否规范）

https：//www.freecollocation.com/（Online Oxford Collocation Dictionary：英语搭配词典）

http：//www.thesaurus.com/（Thesaurus.com：英语同义词词典，查询英语词语的同义表达，便于变换使用的词语）

除了上述在线词典外，"中国知网"根据自己建设的各种学术期刊中、英论文摘要语料库提供的单词翻译查询功能也非常有用。由于中国学者英语水平参差不齐，该单词翻译查询功能仅供参考，但越是专业的词汇，其译法正确的可能性越大，因为这些译法都是专业人士给出的，不可或缺。网址为：https：//dict.cnki.net

---

① 需要说明的是，网上资源变动很快，本"专题讨论"中提到的各种网上资源仅在本书写作时有效，超过此时间段，这些资源是否能够在相应网址查找到，本书编者不敢保证。最佳的策略还是学会如何利用各种搜索引擎快速、准确地查找到自己需要的资源。

另外还有一个功能类似的网站，网址为：http://www.dictall.com/

当然，还有全国科学技术名词审定委员会主办的"术语在线"，已经审定的专业术语的规范译法都可以在网站上查到。网址为：https://www.termonline.cn/

2. 在线百科全书。科技翻译不仅需要双语技能，也需要各种各样的知识，在线百科全书为科技翻译工作者提供了一个快速获得各种百科知识的渠道。一些常用的在线百科全书如：

https://www.britannica.com/（历史悠久、闻名世界的大英百科全书）

https://www.wikipedia.com/（维基百科：免费的百科全书，知识丰富，涉及几乎所有学科和专业，但由于词条编纂人员良莠不齐，内容不是非常可靠，但涉及科技方面的内容准确度还是比较高）

https://baike.sogou.com（搜狗百科：汉语百科，词条质量良莠不齐，但涉及科技方面的内容准确度尚可）

https://baike.baidu.com（百度百科：汉语百科）

https://www.baike.com（快懂百科：汉语百科）

3. 在线语料库。语料库最初是为了语言学家研究语料、编纂词典等建设的，但建成之后，也成了翻译工作者的重要帮手。其作用主要表现在两个方面：一是为译者研究特定词语的用法提供了大量的语料；二是帮助译者确定某种搭配是否正确。一些可以在线查询的语料库有：

http://www.natcorp.ox.ac.uk/（British National Corpus：著名的英国国家语料库）

https://corpus.byu.edu/coca/（The Corpus of Contemporary American English：当代美国英语语料库）

https://corpus.byu.edu/can/（Corpus of Canadian English：加拿大英语语料库）

https://app.sinica.edu.tw/kiwi/mkiwi/（中国台湾"现代汉语平衡语料库"）

4. 视频资源。这些视频资源散布在国内外各种视频平台上，涉及包括冶金、材料等在内的数个学科。这些资源对于非科技背景的语言专业人员特别有益，可以帮助他们了解、熟悉科技制造、生产或实验流程，掌握相关的科技知识，从而提高科技翻译质量。建议外语专业出身且从事科技翻译的人士，先通过这些视频资源获取必要的学科知识，这种方式高效快捷。

5. 在线课程资源。这些课程资源涵盖了几乎所有的学科，教学资源媲美正规课堂。以冶金为例，仅"国家高等教育智慧教育平台"（https://higher.smartedu.cn/）就有冶金类课程近20门，涉及冶金基本原理、冶金传输、粉末冶金、冶金反应、焊接冶金、冶金资源综合利用、稀贵金属冶金等多个方面，为有志于从事冶金科技翻译的人士提供免费的学习资源。

6. 其他资源。这些资源格式多样，有纯文本型、文本+图片型，还有文本+图片+视频型，散落在众多网站中，大多需要通过搜索引擎查找。译者必须熟练掌握常用搜索引擎的使用方法和搜索技巧，准确判断信息的真伪，从而辅助翻译活动的开展。比如，百度文库中有冶金词汇的文档，虽然内容可能有重复，但可靠性较高；微信公众号里有钢铁冶金全流程动画，知乎上有介绍钢铁生产工艺流程的文章。建议有志于从事科技翻译的人士，在学习外语的同时，充分利用这些资源，提高自身的科技素养。

翻译过程中，网上资源有时可以直接帮助译者找到某个科技术语的译法或确认某种译法是否正确无误。以下是几个具体例子：

【例一】

In many shops residual slag is blown with nitrogen to coat the barrel and trunion areas of the vessel. This process is known as "slag splashing".

翻译此句时，编者还没有购买徐树德等编著的《英汉钢铁冶金技术详解词典》，不知道 slag splashing 如何翻译，也不太懂 residual slag is blown with nitrogen to coat the barrel and trunion areas of the vessel 是为了什么，于是以"残余炉渣吹氮"作为关键词，在搜狗搜索引擎中搜索。阅读相关文章后，知道了用氮气将残余炉渣吹向炉壁，是为了保护炉壁上的耐火材料，防止侵蚀，延长其使用寿命。后来，编者又登录"术语在线"网站，查找到 slag splashing 的规范译法是"溅渣护炉"。

【例二】

Coke strength was evaluated using the drum index (DI), the coke strength after reaction (CSR), and the uniaxial compression test with the furnace.

可以看出，原句学术性较强，有四个科技术语。对于这样的句子，尽管目前机器翻译的准确率已经很高了，但术语译法仍需校核。查询《英汉钢铁冶金技术详解词典》可知，drum index (DI) 可译为"转鼓指数"。通过在必应搜索引擎中搜索，可知 uniaxial compression test with the furnace 是一种实验室常用的测定焦炭强度的方法，可译为"炉内单轴压缩试验法"。因此，原句可译成：

采用转鼓指数(DI)、反应后焦炭强度(CSR)和炉内单轴压缩试验对焦炭强度进行了评价。

【例三】

编者在翻译时遇到 American Geosciences Institute (AGI)这个专有名词，不确定这里的 Institute 应该如何翻译，因为其对应的汉语有"研究所""研究院""学院""学会""协会"等。于是，编者登录 American Geosciences Institute 官方网站，发现有如下介绍文字：

The American Geosciences Institute is a nonprofit federation of geoscientific and professional associations that represents more than 250,000 geologists, geophysicists, and other earth scientists. Founded in 1948, AGI provides information services to geoscientists, serves as a voice of shared interests in our profession, plays a major role in strengthening geoscience education, and strives to increase public awareness of the vital role the geosciences play in society's use of resources, resilience to natural hazards, and the health of the environment. (https://www.americangeosciences.org)

既然AGI是一个非营利性的地质科学和专业协会联合会，那么这里institute只有"协会"和"学会"两种可能的译法。通过搜索发现，AGI应翻译成"美国地质协会"，而"美国地质学会"对应的组织是"Geological Society of America (GSA)"。

【例四】
在翻译"GIS设备在电力系统输配电环节具有重要作用，因其占地面积小、可靠性高等优点，目前已在国内电力企业中广泛使用"时，编者不确定如何翻译"占地面积小"。碰巧登录了西门子公司网站，发现其在介绍Gas-insulated switchgear 8DA10 (single-busbar)设备时，谈到其优点有如下：

Up to 40.5 kV, 40 kA (3s), 5,000 A busbar, 2,500 A feeder
Metal-enclosed
Single-busbar System (8DA10)
Gas-insulated
Hermetically enclosed
Factory-assembled, type-tested switchgear according IEC 62 271-200
Customer's Benefit:
—Security of Operation, Reliability
—Personal Safety
—Environmental Independence
—Compactness
—Maintenance-free Design
—Cost-efficient, Ecological

其中，compactness正是"占地面积小"的意思，非常合适，因此采用了这一译法。
西门子公司介绍其设备的英语文字被称为汉语原文的可比文本(Comparable Text)。所谓可比文本，就是目的语中与原文在话题、写作目的、使用环境等方面相似的文本，只不过是用目的语撰写的。比如，若将一份扫描透射电子全息显微镜的汉语使用说明译成英语，那么类似显微镜的英语使用说明就是可比文本。很显然，可比文本包含大量的术语、表达式，可直接用在翻译中。

可比文本能够给译者巨大的帮助。进行翻译前，有意识地阅读一些目的语中的可比文本，能让译者熟悉相关专业和学科，获取大量词汇（包括术语）和表达式，了解文本风格，感知目的语读者的信息需求和美学需求。

总之，科技翻译应充分利用多模态网络资源。科技文本科学性为主，艺术性为辅，可以大胆利用机器翻译，译后人工编辑，校对术语、句型乃至文本的衔接与连贯，起到事半功倍的效果。机器翻译引擎在不断更新升级，科技文本机器翻译的准确率已经非常高，译后认真编辑、审校即可。不过，即使如此，掌握网络资源的利用技巧和方法，对于一个成功的科技翻译工作者来说，仍然是非常必要的。

◎ 练习四　请将下面的句子翻译成汉语。

1. The dynamic development of coke production in Poland requires the application of effective methods for the treatment of coking waste-water.

2. In general, when a load is applied to a material, elongation occurs, creep deformation is likely to occur at high temperatures, and its strain increases until fracture.

3. Very high concentrations of pollutants and their toxicity lead to the necessity of multi-stage treatment before being discharged into water, ground, and the sewage system of an external entity.

4. Coke reacts in a blast furnace, and its strength decreases. In other words, coke is exposed to high-temperature CO_2 and degraded due to the chemical reaction. For industrial evaluation, coke strength after reaction (CSR) is also used as a strength index.

5. On the basis of the examinations of porous structure, it can be concluded that the selected dust sorbents can be used as effective adsorbents for coking wastewater pretreatment.

6. The scope of the study included the examinations of sorption properties of selected sorbents (coal dust, coke dust, biochar), physicochemical tests of coking wastewater after biological treatment, as well as the examinations aimed to determine the dose of adsorbents and time needed to establish the equilibrium state of the process.

7. To measure the coke strength after reaction (CSR), 200 g of coke with a diameter of 20±1 mm was reacted with CO_2 at the temperature of 1,100℃ for 2 hours. The flow rate of CO was 5 NL/min., 1,100℃ for 2 hours.

8. Adsorption techniques play an important role in water and wastewater treatment due to their very high efficiency and lack of selectivity in removing various harmful organic compounds.

9. In the coke-making process, a higher ratio of high-sulfur coal in the coal blend will not only increase the sulfur content in resultant coke, leading to a change of blast furnace operation and unqualified iron, but also increase the load of desulfurization process.

10. In the present study, to investigate the difference in coke strength between at room temperature and at a high temperature, we measured the coke strength in high-temperature CO_2 reaction atmosphere.

11. Coal properties, minerals and sulfur forms in coal, and pyrolysis conditions such as temperature, atmosphere, and the mass transfer, are the main factors that significantly influence the sulfur transformation behavior during coal pyrolysis.

12. Since the CO_2 reaction decreased the weight of coke, the weight of the sample would be decreased. Unfortunately, the weight of fine particles generated by the fracture and chemical reactions cannot be measured.

第五课　高炉是如何运转的

> （一）
> 吞吐八荒气势雄，巍然矗立傲苍穹。
> 胸中烈烈一团火，化作霞光万里红。
> （二）
> 一生功业在熔融，去伪存真求大同。
> 消得云烟千万缕，不教块垒积胸中。
>
> ——刘耀业《高炉颂》（二首）①

一、科技术语

beneficiate（Mining & Quarrying）To process (ores) through reduction.（选矿）

dolomite A magnesia-rich sedimentary rock resembling limestone.（白云石）

flux A mineral added to the metals in a furnace to promote fusing or to prevent the formation of oxides.（熔剂）

gangue The valueless rock or aggregates of minerals in an ore.（脉石）

impurities Worthless or dangerous material that should be removed.（杂质）

inclined skip hoist A skip hoist that operates on steeply inclined rails placed on a mine pit slope or wall.（爬式加料机）

pellet A piece of ore that is obtained from fine, dustlike ore or from finely ground concentrates, which is very firm spherical lump that range in diameter from 2-3 mm to 30 mm.（球团矿）

permeability The capability of a porous rock or sediment to permit the flow of fluids or gases through its pore spaces.（透气度）

---

① 鞍山钢铁/鞍山市文联"新时代·新鞍钢"主题诗词创作活动作品选登[EB/OL].搜狐网，https://roll.sohu.com/a/576439019_121123735[2022-8-13].

refractory brick Brick that can withstand high temperatures.(耐火砖)

sinter A type of ore material formed from fusible ore powder by holding the pressed powder at a temperature just below its melting point for a period of time so that the particles are fused (sintered) together, but the mass, as a whole, does not melt.(烧结矿)

slag Also called "cinder", the fused material formed during the smelting or refining of metals by combining the flux with gangue, impurities in the metal, etc. It usually consists of a mixture of silicates with calcium, phosphorus.(炉渣)

taphole A hole in a furnace or ladle through which molten metal is tapped.(开铁口)

二、英语原文

How a Blast Furnace Works

1 The purpose of a blast furnace is to chemically reduce and physically convert iron oxides into liquid iron called "hot metal". The blast furnace is a huge, steel stack lined with refractory brick, where iron ore, coke and limestone are dumped into the top, and preheated air is blown into the bottom. The raw materials require 6 to 8 hours to descend to the bottom of the furnace where they become the final product of liquid slag and liquid iron. These liquid products are drained from the furnace at regular intervals. The hot air that was blown into the bottom of the furnace ascends to the top in 6 to 8 seconds after going through numerous chemical reactions. Once a blast furnace is started it will continuously run for four to ten years with only short stops to perform planned maintenance.

The Process

2 Iron oxides can come to the blast furnace plant in the form of raw ore, pellets or sinter. The raw ore is removed from the earth and sized into pieces that range from 0.5 to 1.5 inches. This ore is either Hematite (Fe_2O_3) or Magnetite (Fe_3O_4) and the iron content ranges from 50% to 70%. This iron rich ore can be charged directly into a blast furnace without any further processing. Iron ore that contains a lower iron content must be processed or beneficiated to increase its iron content. Pellets are produced from this lower iron content ore. This ore is crushed and ground into a powder so the waste material called gangue can be removed. The remaining iron-rich powder is rolled into balls and fired in a furnace to produce strong, marble-sized pellets that contain 60% to 65% iron. Sinter is produced from fine raw ore, small coke,

sand-sized limestone and numerous other steel plant waste materials that contain some iron. These fine materials are proportioned to obtain a desired product chemistry then mixed together. This raw material mix is then placed on a sintering strand, which is similar to a steel conveyor belt, where it is ignited by gas fired furnace and fused by the heat from the coke fines into larger size pieces that are from 0.5 to 2.0 inches. The iron ore, pellets and sinter then become the liquid iron produced in the blast furnace with any of their remaining impurities going to the liquid slag.

3 The coke is produced from a mixture of coals. The coal is crushed and ground into a powder and then charged into an oven. As the oven is heated the coal is cooked so most of the volatile matter such as oil and tar are removed. The cooked coal, called coke, is removed from the oven after 18 to 24 hours of reaction time. The coke is cooled and screened into pieces ranging from one inch to four inches. The coke contains 90 to 93% carbon, some ash and sulfur but compared to raw coal is very strong. The strong pieces of coke with a high energy value provide permeability, heat and gases which are required to reduce and melt the iron ore, pellets and sinter.

4 The final raw material in the ironmaking process is limestone. The limestone is removed from the earth by blasting with explosives. It is then crushed and screened to a size that ranges from 0.5 inch to 1.5 inch to become blast furnace flux. This flux can be pure high calcium limestone, dolomitic limestone containing magnesia or a blend of the two types of limestone.

5 Since the limestone is melted to become the slag which removes sulfur and other impurities, the blast furnace operator may blend the different stones to produce the desired slag chemistry and create optimum slag properties such as a low melting point and a high fluidity.

6 All of the raw materials are stored in an ore field and transferred to the stockhouse before charging. Once these materials are charged into the furnace top, they go through numerous chemical and physical reactions while descending to the bottom of the furnace.

7 The iron ore, pellets and sinter are reduced which simply means the oxygen in the iron oxides is removed by a series of chemical reactions. These reactions occur as follows:

1) $3Fe_2O_3 + CO = CO_2 + 2Fe_3O_4$ Begins at 850 °F

2) $Fe_3O_4 + CO = CO_2 + 3FeO$ Begins at 1,100 °F

3) $FeO + CO = CO_2 + Fe$
   $FeO + C = CO + Fe$ Begins at 1,300 °F

8 At the same time the iron oxides are going through these purifying reactions, they are also beginning to soften then melt and finally trickle as liquid iron through the coke to the bottom of the furnace.

9 The coke descends to the bottom of the furnace to the level where the preheated air or hot blast enters the blast furnace. The coke is ignited by this hot blast and immediately reacts to generate heat as follows:

$$C + O_2 = CO_2 + Heat$$

10 Since the reaction takes place in the presence of excess carbon at a high temperature the carbon dioxide is reduced to carbon monoxide as follows:

$$CO_2 + C = 2CO$$

11 The product of this reaction, carbon monoxide, is necessary to reduce the iron ore as seen in the previous iron oxide reactions.

12 The limestone descends in the blast furnace and remains a solid while going through its first reaction as follows:

$$CaCO_3 = CaO + CO_2$$

13 This reaction requires energy and starts at about 1600°F. The CaO formed from this reaction is used to remove sulfur from the iron which is necessary before the hot metal becomes steel. This sulfur removing reaction is:

$$FeS + CaO + C = CaS + FeO + CO$$

14 The CaS becomes part of the slag. The slag is also formed from any remaining Silica (SiO_2), Alumina (Al_2O_3), Magnesia (MgO) or Calcia (CaO) that entered with the iron ore, pellets, sinter or coke. The liquid slag then trickles through the coke bed to the bottom of the furnace where it floats on top of the liquid iron since it is less dense.

15 Another product of the ironmaking process, in addition to molten iron and slag, is hot dirty gases. These gases exit the top of the blast furnace and proceed through gas cleaning equipment where particulate matter is removed from the gas and the gas is cooled. This gas has a considerable energy value so it is burned as a fuel in the "hot blast stoves" which are used to preheat the air entering the blast furnace to become "hot blast". Any of the gas not burned in the stoves is sent to the boiler house and is used to generate steam which turns a turbo blower that generates the compressed air known as "cold blast" that comes to the stoves.

16 In summary, the blast furnace is a counter-current reactor where solids descend and gases ascend. In this reactor there are numerous chemical and physical reactions that produce the desired final product which is hot metal. A typical hot metal chemistry follows:

The Blast Furnace Plant

  Iron (Fe)   = 93.5% to 95.0%
  Silicon (Si)   = 0.30% to 0.90%
  Sulfur (S)   = 0.025% to 0.050%

Manganese (Mn) = 0.55% to 0.75%
Phosphorus (P) = 0.03% to 0.09%
Titanium (Ti) = 0.02% to 0.06%
Carbon (C) = 4.1% to 4.4%

17 Now that we have completed a description of the ironmaking process, let's review the physical equipment comprising the blast furnace plant.

Conclusion

18 The blast furnace is the first step in producing steel from iron oxides. The first blast furnaces appeared in the 14th century and produced one ton per day. Blast furnace equipment is in continuous evolution and modern, giant furnaces produce 13,000 tons per day. Even though equipment is improved and higher production rates can be achieved, the processes inside the blast furnace remain the same. Blast furnaces will survive into the next millenium because the larger, and more efficient furnaces can produce hot metal at costs competitive with other iron making technologies.

三、分析与讲解

高炉炼铁是钢铁冶金工业的核心流程。高炉炼铁工艺成熟，技术经济指标良好，生产量大，劳动生产效率高，因此高炉炼铁生产的铁占世界铁总产量的绝大部分。

我国是最早使用高炉的国家之一。西汉时期的熔炉可以算得上是世界上最古老的高炉。在我国出土的公元前5世纪文物中发现的铸铁表明，当时熔炼技术已经实用化。

高炉炼铁的原理其实很简单：在高温的作用下，焦炭中的碳原子与铁矿石中的氧原子结合，从而还原出铁矿石中的铁。但具体的高炉生产涉及诸多设备和多个流程，具体分析如下：

文章标题：可以译为"高炉是如何运转的"，也可以译为"高炉是如何工作的"。"运转"更正式一些，也更具技术气息。

第1段：(1)第1句中的 physically 当然来自 physical，后者又来自 physics。查英英词典，physical 有如下释义：

adj. ① a) Of or relating to the body. See Synonyms at bodily. b) Having a physiological basis or origin: a physical craving for an addictive drug. c) Involving sexual interest or activity: a physical attraction; physical intimacy.

② a) Involving or characterized by vigorous or forceful bodily activity: physical aggression; a fast and physical dance performance. b) Slang Involving or characterized by violence：" A real cop would get physical" (TV Guide) .

③ Of or relating to material things：a wall that formed a physical barrier; the physical environment.

④ Of or relating to matter and energy or the sciences dealing with them, especially physics.

n. A physical examination.

陆谷孙主编的《英汉大词典》对于 physical 有如下解释：

adv. ①按自然规律。②(口语)完全地，全然。常用 physically impossible：It's physically impossible for me to finish this by Christmas. 要我在圣诞节之前完成这项工作是完全不可能的。③身体上。He looked physically fit. 他看上去身体很健康。④实际上，真正地。Some religions suppose that there is a Heaven physically above the earth. 有些宗教认为地球之上真的有一个天国。

具体到这里的 physically，有两种理解：一是"物理上"，二是"实际上、真正地"。从表达上来说，两种理解都没问题，但从逻辑上来说，既然是 chemically reduce(将铁氧化物转化为液态铁)，应该不会是"物理上" convert iron oxides into liquid iron，应作后一种理解。为保险起见，可以不翻译，这并不影响信息的传递。

(2)翻译第 2 句时，需要注意不能完全按照原句的结构和顺序翻译，而要变换句型："高炉体型庞大，炉身为钢制，内衬耐火砖……"这样的表达，远比"高炉是一个巨大的、衬有耐火砖钢制炉身"要好。(3)由于 6 to 8 hours 和 6 to 8 seconds 中的数字都不大，为了保持体例的一致，建议翻译时使用汉字数字"六到八个小时"和"六到八秒"。(4) ascends to the top in 6 to 8 seconds after going through numerous chemical reactions 当然可以翻译成"经过很多化学反应后在六到八秒内上升到炉顶"，但不如按照原句的顺序翻译成"在六到八秒内上升到炉顶，其间经过了很多化学反应"读起来更加自然流畅。

科技英语中包含大量的阿拉伯数字。译成汉语时，是采用原来的阿拉伯数字，还是改用汉字数字，要根据《出版物上数字用法》国家标准(GB/T 15835-2011)来决定，不能随意选择。

第 1、2 段间小标题：这一小节主要讨论高炉如何炼铁，所以小标题可译为"炼铁过程"。

第 2 段：(1)汉语中一般不说"高炉工厂"，所以第 1 句中的 the blast furnace plant 可以译成"高炉所在地"。英语和汉语词汇有不同的搭配模式，不仅汉译英需要注意搭配，英译汉同样需要注意汉语词语间的搭配。(2)第 2 句中 sized into pieces that range from 0.5 to 1.5 inches 中的 pieces 很难译，可以直接不译："被加工成 0.5—1.5 英寸(1.27—3.81 厘米)大小。"翻译时会遇到某些词语很难翻译的情况，但不译并不影响意思的表达，这时可以选择不译。(3)将被动语态的句子译成无主句是一种惯用的翻译方法。例如，将 This iron ore is crushed and ground into a powder so that the waste material called gangue can be removed 译成"将矿石破碎磨成粉，以便去除被称为脉石的无用物质"。(4) The remaining iron-rich powder is rolled into balls and fired in a furnace to produce strong, marble-sized pellets that contain 60% to 65% iron 一句带有定语从句，可以通过重复定语从句修饰的名词来翻译："剩余的含铁量高的粉末被滚动造球，放炉中焙烧，形成弹珠大小、强度高的球团。球团含铁"在60%—65%"。这种通过重复先行词翻译定语从句的方法很常用。

科技英语中常常会碰到英制度量衡单位，如英寸(inch)、英尺(foot，复数为 feet)、磅(pound)、英里(mile)、盎司(ounce)、加仑(gallon)等。仅仅将这些单位翻译成汉语，虽然也算翻译，但这样的翻译不能满足大部分读者的需求，毕竟大多数汉语读者对这些度量衡单位没有多少概念。因此，最好的策略是：翻译这些英制单位，同时用括号给出换算成公制单位的值。比如"0.5—1.5英寸"后面就应给出用公制厘米来换算的值"(1.27—3.81厘米)"。其他地方也是如此。

第3段：(1)若将第1句译成"焦炭生产自混合的煤炭"，那么第2句翻译时就需要在主语"煤炭"前增加"混合好的"，以便衔接前句。(2)第3、4句中的 cooked 不好翻译。第3句中的 cooked 意为"被加热"，也就是"受热"。第4句中的 cooked 则表示"受热过程已经结束了"，由于汉语中常用"炼焦"一词，所以可以译成"炼好的"。(3)科技文本里，形容词的翻译要注意用词的专业性。如将 strong 译为"高强度的"比"结实的"更准确；用"高热值"来译"with a high energy value"也比"具有高能量价值的"更简洁。(4)最后一句的翻译，需要注意用词的技术性，将 strong 译成"高强度"，将 a high energy value 译成"高热值"。

第4段：第2句中的 from the earth 不能译为"从地下"，因为有些石灰石矿位于地表之上，所以应直接按照原词组译成"从地球中"。

第6段：虽然查"术语在线"可以发现，ore field 对应的表达是"矿田"，但"矿田"条目下给出的定义是："由一系列矿床组合而成的含矿地区或矿化点、物化探异常最集中的地区"，显然这个定义并不适用于此处。进一步查阅《英汉钢铁冶金技术详解词典》，可以发现"贮矿场"才是更准确的译法。

虽然"术语在线"是权威机构为规范、统一汉语科技术语使用而建设的网站，但其提供的译法并不能机械地套用，要看"术语在线"提供的译法是否符合词语所在的使用语境。如不符合，则应查询其他专业词典，再作选择。

第7段：翻译℉时，不仅要给出华氏度，还需要用括号带出转换为摄氏度的值，以便读者理解。其他地方也是如此。

第8段：(1)查阅《英汉钢铁冶金技术详解词典》以及在线词典等，purifying reactions 均被译为"净化反应"或"精炼反应"，但这两个词语都不适合描述铁氧化物在高炉中经历的化学变化，所以不妨译成"熔炼反应"。(2) trickle through the coke to the bottom of the furnace 不好译。高炉中的焦炭不仅为铁矿石还原提供碳源，同时起到支撑的作用，防止铁矿石过快下降到炉底而无法充分还原，所以 through the coke 应译为"从焦炭缝隙中"。至于 trickle，不仅有"滴"的意思，也有"流"的意思。所以，整个表达可译为"从焦炭缝隙中滴流到炉底"。

第 9 段：descends to 在这里不是"下降到"，而是"朝……下降"或"向……下降"，因为后面还有一个 to the level where the preheated air or hot blast enters the blast furnace。

第 16 段：A typical hot metal chemistry 中的 chemistry 当然不能译成"化学"，而要译成"化学成分"。

第 17 段：(1) review 这里不是"复习"，而是"仔细检视"的意思，可以简单地译为"看一看"。(2) 这里的 blast furnace plant 不能译成"高炉所在地"了，而应译成"高炉车间"，接下来的小标题也是如此。(3) physical equipment 中的 physical 不是"物理的"，而是"看得见的"，或者说"有形的"。equipment 不仅包括"设备"，也包括高炉车间的"设施"，例如第 18 段第 1 句提到的 ore storage 就是"设施"。所以 physical equipment 应译为"有形设施和设备"。选词永远是翻译过程的一项重要工作。

第 17、18 段间小标题：Conclusion 不能译为"总结"，因为第 18 段的内容不是总结，而是一段结束的话语，所以可以译成"结语"。

第 18 段：(1) 原文说 modern, giant furnaces produce 13,000 tons per day。事实上，我国沙钢集团的一座高炉容积高于 5800 立方米，年产生铁 500 多万吨，是世界上最大、最先进的高炉。这座高炉的优势在于它拥有世界上最大的鼓风机，重量达到 123 吨，输出的功率有 6 万千瓦，风压达到了惊人的 52 万帕，是十二级飓风的一千倍！(2) 倒数第二句，原文从句中用的是一般现在时，译成汉语时，不妨改成进行时："虽然设备在改进，生产速度在提高，……"

四、参考译文

高炉是如何运转的

1 高炉是用来化学还原铁氧化物并将铁氧化物转换为被称为"铁水"的液态铁的。高炉体型庞大，炉身为钢制，内衬耐火砖。从炉顶装入铁矿石、焦炭和石灰石，从炉底吹入预热空气。原材料经过六到八个小时下降到炉底，变成最终产品——液态炉渣和液态铁。这些液态产品定期从高炉排出。热空气从炉底吹入，在六到八秒内上升到炉顶，其间经过了很多化学反应。高炉一旦开始生产，就会连续运转四到十年，只是间隔短时停产，以便按计划开展检修。

炼铁过程

2　铁氧化物以原矿、球团矿或烧结矿的形式被运到高炉所在地。原矿从地球开采出来，被加工成0.5—1.5英寸(1.27—3.81厘米)大小。原矿或者是赤铁矿(Fe_2O_3)或者是磁铁矿(Fe_3O_4)，含铁量在50%—70%。这些含铁量高的矿石可以不经处理直接装入高炉。而含铁量较低的矿石则必须经过选矿等处理增加其含铁量。球团矿就是来自于这种含铁量低的矿石。将矿石破碎磨成粉，以便去除被称为脉石的无用物质。剩余的含铁量高的粉末被滚动造球，放炉中焙烧，形成弹珠大小、强度高的球团。球团含铁在60%~65%。烧结矿生产用到的是原矿细粉、小粒度焦炭、沙粒大小的石灰石和很多含铁的钢厂废弃材料。这些材料细粒按一定比例配好，以便产品能获得期望的化学性质，然后混匀。将混匀的原材料放置在外形类似钢制皮带运输机的烧结机上，点燃煤气炉，焦炭粉末燃烧的热量将原材料熔化，形成颗粒较大、尺寸在0.5—2.0英寸(1.27—5.08厘米)间的烧结矿。铁矿石、球团矿和烧结矿在高炉里变成铁水，剩余杂质进入液态炉渣。

3　焦炭生产自混合的煤炭。将混合好的煤炭破碎，磨成粉末，送入焦炉内。给焦炉加热，煤炭受热，大部分挥发性物质被除去，例如油和焦油。经过18—24小时的反应后，从焦炉移出"炼"好的煤炭，即焦炭。将焦炭冷却，筛分出1—4英寸(2.54—10.16厘米)不等的焦炭。焦炭含90%—93%的碳、一些灰分和硫。与原煤比，焦炭强度要大得多。高强度、高热值的焦炭能给熔化和还原铁矿石、球团矿和烧结矿提供所需的透气性、热量和气体。

4　炼铁的最后一种原材料是石灰石。石灰石是通过炸药爆炸的方式从地球中开采出来的。然后将其破碎、过筛，得到0.5—1.5英寸(1.27—3.81厘米)的石灰石，成为高炉助熔剂。助熔剂可以是纯高钙石灰石，也可以是含镁的白云质灰岩，或两者的混合。

5　因为石灰石熔化后成为炉渣，可以去除硫和其他杂质，因此高炉操作人员会将不同种类的石头混合，带给炉渣期望的化学性质，产生的炉渣具有最佳的特性，如低熔点、高流动性。

6　所有这些原材料都存放在贮矿场，在装料前被运到料仓。这些材料从炉顶装入高炉后，经过多次物理和化学反应，与此同时，材料渐渐向炉底下降。

7　铁矿石、球团矿和烧结矿被还原，即铁氧化物中的氧被一系列化学反应去除。这些反应有：

1) $3Fe_2O_3+CO=CO_2+2Fe_3O_4$ 反应开始于850℉(454.4℃)

2) $Fe_3O_4+CO=CO_2+3FeO$ 反应开始于1100℉(593.3℃)

3) $FeO+CO=CO_2+Fe$ 或 $FeO+C=CO+Fe$ 反应开始于1300℉(704.4℃)

8　铁氧化物经历这些熔炼反应的同时，也开始变软、熔化，最后变成铁水从焦炭缝隙中滴流到炉底。

9 焦炭也朝炉底下降，降至预热空气或热风进入高炉的高度，被热风点燃，立刻发生反应，产生热量，如下：

$$C+O_2=CO_2+热量$$

10 因为反应是在高温中发生的，且有多余的碳，因此二氧化碳被还原成一氧化碳，如下：

$$CO_2+C=2CO$$

11 反应生成物一氧化碳为还原铁矿石所必需，这从前面给出的铁氧化物反应式可看出。

12 石灰石在高炉中下降时，保持固体状态，同时经历第一个反应，如下：

$$CaCO_3=CaO+CO_2$$

13 这个反应需要热量，开始于 1600℉（871.1℃）。这个反应生成的 CaO 用于去除铁中的硫。铁水在变成钢之前，这一步是必要的。除硫反应如下：

$$FeS+CaO+C=CaS+FeO+CO$$

14 反应后，CaS 成为了炉渣的一部分。随铁矿石、球团矿、烧结矿或焦炭一起进入高炉残留在高炉中的二氧化硅（SiO_2）、氧化铝（Al_2O_3）、氧化镁（MgO）或氧化钙（CaO）也会形成炉渣。液态炉渣穿过焦床滴流到炉底，因为密度较低，漂浮在铁水之上。

15 除铁水和炉渣外，炼铁过程的另一个产品是灼热的、肮脏的炉气。这些炉气从炉顶排出，经由炉气清洁设备，除去其中的微粒，然后冷却。这种炉气具有相当的热值，因此被送往热风炉作为燃料燃烧，用来预热进入高炉的空气，将这些空气变成"热风"。热风炉没有用上的炉气被送往锅炉房，用来生产蒸汽，推动涡轮鼓风机，生产被称为"冷风"的压缩空气，用于热风炉。

16 概括起来，高炉就是一个逆流反应器，固体物料下降，气体上升。在这个反应器里，发生很多物理化学反应，最后生产出期望的产品，即铁水。典型铁水的化学成分如下：

铁（Fe）　　= 93.5%—95.0%

硅（Si）　　= 0.30%—0.90%

硫（S）　　= 0.025%—0.050%

锰（Mn）　　= 0.55%—0.75%

磷（P）　　= 0.03%—0.09%

钛（Ti）　　= 0.02%—0.06%

碳（C）　　= 4.1%—4.4%

17 既然我们已经完整地描述了炼铁的过程，现在就让我们看一看构成高炉车间的设施和设备。

高炉车间

(1) casthouse 出铁场	(2) gas cooler 炉气冷却器
(3) dustcatcher 集尘器	(4) Venturi Scrubber 文氏管洗净器
(5) bells 料钟	(6) receiving hopper 受料料斗
(7) downcomer 下降管	(8) bleeder valves 分压阀
(9) offtake 排气管	(10) uptakes 上升管
(11) stack 烟囱	(12) hot blast stove 热风炉
(13) hot blast main 热风主管	(14) cold blast main 冷风主管
(15) mixer line 混风管	(16) inclined skip bridge 上料斜桥
(17) stockhouse/hiline 料房/栈桥料仓	(18) storage bins 储料仓/槽
(19) skip car 上料车	(20) hoist house 绞车房
(21) cinder notch 出渣口	(22) iron notch 出铁口
(23) taphole drill 开铁口钻头/开口机	(24) slag runners 渣沟
(25) slag pots 渣罐	(26) ladles 铁水包
(27) iron runners 铁沟	(28) trough 主铁沟
(29) mudgun 泥炮	(30) tuyeres 风口
(31) bustle pipe 热风围管	

  18 第一项设施是矿石堆场。矿石堆场也可以是船舶卸货的矿石码头。存放在堆场里的原材料包括原矿石、几种球团矿、烧结矿、石灰石或混合助熔剂，可能还有焦炭。这些材料通过配备有抓斗的桥式起重机或皮带运输机被运送到料房/栈桥料仓综合体(17)。也可通过轨道式料车将材料运送至料房/栈桥料仓或从桥式起重机运送到被称为"矿石输送车"的自行式轨道车上。每种矿石、球团矿、烧结矿、焦炭和石灰石都被装进单独的"储料仓/槽"(18)。不同的原材料要进行称重，根据某一配料设计，决定材料用量，以便铁水和炉渣的化学性质符合期望。材料称重是在储料仓下通过在轨道上安装的磅秤车或给皮带运输机供料的计算机控制的称重料斗来实现。称量好的材料被装进轨道上行驶的上料车(19)、迎着倾斜的料桥向上被倒进炉顶的受料料斗(6)。提升上料车的缆索由位于绞车房(20)的大型绞车推动。有些现代化的高炉采用从料房到炉顶完全自动化的传送机完成这一工作。

  19 在炉顶，材料短暂停留，直到通常由某些种类的含铁矿物(矿石、球团或烧结矿)、焦炭和助熔剂(石灰石)构成的炉料全部聚齐。高炉操作工根据严格控制炉气流动和高炉内化学反应的要求，确定具体的布料顺序。材料经过圆锥形料钟(5)的两段布入高炉。料钟负责沿炉喉的圆周将原材料均匀地布入高炉，同时将炉气密封在高炉内。有些现代化高炉没有料钟，代之以两到三个闭锁料仓，闭锁料仓将原材料倒进旋转溜槽，旋转溜槽可以改变角度，以便更灵活准确地在炉内布料。

20　炉顶还有四个上升管(10)，灼热、肮脏的炉气借此从炉顶逸出。两个上升管汇进一个排气管，炉气经上升管流动到排气管(9)。两个排气管汇进下降管(7)。在上升管的最顶端，有释放炉气保护炉顶免遭突然增加的炉气压力损坏的分压阀(8)。炉气在下降管下降至集尘器，在集尘器里，炉气中的粗颗粒下沉聚集，(达一定量时)由火车或卡车运出处理。然后炉气流经文氏管洗净器(4)除去其中较细的颗粒，最后进入炉气冷却器(2)，通过喷水，降低灼热但却干净的炉气的温度。有些现代化高炉配备有(炉气)洗净冷却一体化装置。至此，洁净、冷却了的炉气可以用于燃烧了。

21　输送洁净炉气的管道通向热风炉(12)。高炉附近通常有一排三到四座圆柱形的热风炉。炉气在热风炉底部燃烧，热气上升，传到炉内的耐火砖。燃烧的产物穿过耐火砖间的通道，流出热风炉，进入一座很高的"烟囱"(11)，所有的热风炉共享一个"烟囱"。

22　涡轮鼓风机产生大量的空气——80000—230000 ft^3/min——流经"冷风主管"(14)，到达热风炉组。冷风进入之前加过热的热风炉，炉内耐火砖里储存的热量传递到"冷风"，形成"热风"。取决于热风炉的设计和内部情况，热风温度在1600°F(871.1℃)到2300°F(1260℃)之间。经过加热的空气离开热风炉，进入"热风主管"(13)，热风主管通向高炉。冷风主管和热风主管通过一混风管(15)连接，混风管配有一个阀门，控制热风温度，保持温度恒定。热风主管进入一个炸面圈形状的环绕高炉的管道，称为"热风围管"(31)。热风从热风围管通过被称为"风口"(读作"tweers")的喷嘴被输送进高炉。风口沿高炉圆周以相等间隔布置。小型高炉可能会有14个风口，大型高炉的风口可能会达到40个。这些风口为铜制，采用水冷，这是因为风口前的温度可能高达3600°F(1982.2℃)—4200°F(2315.6℃)也可以在风口水平向高炉里喷入石油、焦油、天然气、煤粉和氧气。这些喷入物与碳结合，释放多余热量，为提高生产率所必需。铁水和炉渣经过风口，向炉缸流去，而炉缸就在风口水平之下。

23　在高炉下半部周围，围绕热风围管、风口和出铁出渣设备的是出铁场(1)。炉缸上用来出铁或者说将铁水放出高炉的开口叫"出铁口"(22)。一个安装在旋转基座上被称为"开铁口钻头/开口机"的巨大钻头旋转到出铁口，在耐火黏土塞上钻出一个孔，直至铁水。高炉上另外一个开口，被称为"出渣口"，用于出渣，或紧急情况出铁。一旦出铁口被钻开，铁水和渣就会向下流入一个深沟，称为"主铁沟"(28)。嵌入主铁沟并横跨主铁沟的是一个耐火块，叫作"撇渣器"。撇渣器下部有一小孔，铁水流过小孔，越过铁沟坝，进入铁沟。因为炉渣密度比铁小，所以浮在铁水上面，流经主铁沟时，撞上撇渣器，被导入"渣沟"。液态炉渣最后流入"渣罐"(25)或渣坑(表格中未显示)，铁水流入有耐火材料内衬的"铁水包"(26)，铁水包因其外形而被称为"鱼雷罐车"或"潜艇式车"。高炉铁水降低至出铁口水平时，从风口鼓进高炉的某些热风会让铁水从出铁口喷溅出来，这是出铁结束的信号，因此"泥炮"(29)会被移进出铁口。泥炮气缸已提前装满耐火粘土，此时启动，气缸活塞将耐火粘土推进出铁口，阻止铁水流出。出铁结束后，铁水包被送至炼钢厂炼钢，炉渣被运至堆渣场，加工后用于道路填埋或铁路道碴。然后清理出铁场，准备迎接45分钟到2小时后的下一次出铁。现代化的大型高炉可能会有高达四个出铁口和两个出铁场。出铁速度必须与原材料装料速度及铁/渣生成速度相同，以便液面保持在炉缸以内、风口的高度以下。如果液面高过风口，则可能会烧毁铜铸件，损坏炉衬。

结语

24　高炉只是用铁氧化物炼钢的第一步。第一座高炉出现在14世纪，每天产铁1吨。高炉设备不断发展，现代化巨型高炉每天可产铁13000吨。虽然设备在改进，生产速度在提高，但高炉炼铁流程并没有发生变化。下一个千年，高炉仍将存在，因为与其他炼铁技术相比，大型、高效的高炉生产铁水的成本更有竞争力。

五、专题讨论

科技翻译中汉语无主句的使用

一、汉语无主句的概念

汉语存在大量没有明确主语的句子。这些句子要么在主语位置上没有任何成分，要么主语位置上的成分只是用来提起一个话题，而谓语部分是对该话题的评论。这类句子缺乏英语所强调的语法功能(主谓兼备)和逻辑功能(施动和受动关系)，是汉语中常见的句法现象，被称为"无主句"。例如：

> 下雨了。
> 吃一堑，长一智。
> 远处传来一阵枪声。
> 刚才有你的一个电话。

吕叔湘先生最早提出汉语是"话题凸显"的语言，而英语是"主语凸显"的语言这一观点。汉语句子中，主语位置上的成分不仅表示动作的执行，有时还表示工具、原因、比较的对象，变化的结果和关涉的问题等。[①] 相较而言，英语句子是围绕"主语+谓语"这个核

---

① 吕叔湘. 汉语语法分析问题[M]. 北京：商务印书馆，1979：72.

心轴建构的，主语和谓语前后可以添加很多定语、状语、补语等成分。英语句子的主语一般分为心理主语、逻辑主语和语法主语。① 其中，语法主语和逻辑主语是语言学习者非常熟悉。前者可以用来判断句子的真假值——没有主语会导致句子不成立，并决定句子的语法特征（如人称和数一致），对应的功能为 subject（主语）。后者为动作的执行者，其对应的功能为 actor（施事）。英语和汉语在主语位置上的结构性差异导致我们在进行科技英语汉译时，常需要使用汉语的无主句，以满足读者的心理期待。例如：

Attention must be paid to the working temperature of the machine.
应当注意机器的工作温度。
It is not recommended to keep ECMO intubation.
不建议保留 ECMO 插管。
Review of his past history was negative except for occasional mild upper respiratory infections.
回顾过去病史，除偶有轻微上呼吸道感染外，其余皆正常。

这里三个例子均采用了汉语无主句进行翻译。虽然译句在句法结构上不同于原句，但都符合汉语的表达习惯。英汉科技翻译时，应充分利用汉语无主句的表达优势，创造出准确、地道的译文。

二、汉语无主句在英汉科技翻译中的应用

文学翻译也好，科技翻译也罢，只要是英汉翻译，通常需要将重形合和逻辑关系的英语表达转换成重意合和主题关系的中文表达。在这个过程中，大胆使用汉语无主句，可以有效地帮助译者完成这种转换。

汉语无主句常可以用来翻译四种类型的英语句子：含被动语态的句子，非谓语动词逻辑主语不明的句子，there be 句型，以及 it 引导的句子。

1. 被动语态。科技英语中大量使用被动语态，目的常是强调非灵主语，凸显叙述的客观。翻译时，通常可使用汉语的无主句，以突出英语中聚焦的主题事件。例如：

…where iron ore, coke and limestone are dumped into the top, and preheated air is blown into the bottom.
……从炉顶装入铁矿石、焦炭和石灰石，从炉底吹入预热空气。
This ore is crushed and ground into a powder so the waste material called gangue can

---

① 陈脑冲. 论主语[J]. 外语教学与研究, 1993 (4): 5.

be removed.

将矿石破碎磨成粉，以便去除被称为脉石的无用物质。

上述两个例子中，原句主语 iron ore, coke and limestone 和 This ore 都是受事对象，但施事者并不清楚。译成汉语时，可以将这些表受事对象的词语放在动词宾语(装入铁矿石、焦炭和石灰石)或者介词宾语(将矿石破碎磨成粉)的位置上，而将整个句子译成汉语无主句。这种翻译英语被动语态的方法非常普遍，请再看几个例子：

Amniocentesis was performed at 37 weeks and the lecithin-sphingomyelin ratio was reported as 3∶2, phosphatidylglyceral was present.

孕37周行羊膜穿刺术检查，卵磷脂—鞘磷脂比值报告为3∶2，出现磷脂酰甘油。

A little black spot was seen moving quickly on the screen.

在屏幕上可以看到一个小黑点迅速地移动着。

As the oven is heated the coal is cooked so most of the volatile matter such as oil and tar are removed.

给焦炉加热，煤炭受热，大部分挥发物质被除去，例如油和焦油。

2. 非谓语动词逻辑主语不明的句子。科技英语要在有限的空间内传达足够多的信息，常使用很多结构复杂的非谓语动词。有时，这些非谓语动词的逻辑主语不清楚，翻译时便需要使用汉语的无主句，避免说出不明确或没必要明确的施动者。例如：

Once a blast furnace is started it will continuously run for four to ten years with only short stops to perform planned maintenance.

高炉一旦开始生产，就会连续运转四到十年，只是间隔短时停产，以便按计划开展检修。

The remaining iron-rich powder is rolled into balls and fired in a furnace to produce strong, marble-sized pellets.

将剩余的含铁量高的粉末滚动造球，放炉中焙烧，形成弹珠大小、强度很高的球团。

上述两例中，to perform planned maintenance, to produce strong 和 marble-sized pellets 都是非谓语动词短语，它们的逻辑主语不明确，我们不知道谁计划、实施维护，也不知道谁焙烧以形成这样的球团，且这些信息对于高炉运行和生产球团并不重要，因此翻译时，只需按原句结构译出和非谓语动词相关的内容即可，无须添加主语。

3. There be 句型。科技英语中，有不少 there be 句型，表示存在的现象或者状况。其中，there 是虚词，本身无意义，be 后面是真正的主语。翻译这类句子时，也可以使用汉语无主句。例如：

In this reactor there are numerous chemical and physical reactions that produce the desired final product which is hot metal.

在这个反应器里，发生很多物理化学反应，最后生产出期望的产品，即铁水。

Because there is at this stage no acquired immunity, the bacilli multiply rapidly, eliciting an exudative inflammatory response in which polymorphonuclear leucocytes and monocytes predominate.

因为在此阶段并无后天免疫，杆菌繁殖迅速，从而引起渗出性炎症反应，其中以多形核白细胞和单核白细胞居支配地位。

这两个例子中，there be 句型所表达的存在概念和汉语中的"有……""没有……"等引出的无主句所表达的存在概念相同，所以可以使用汉语无主句来翻译。同样 there seems、there appears 等类的句子也可以译成汉语无主句。例如：

There seem to be some problems in the automated control system.
在自动化控制系统方面似乎有问题。
There appears to be a reflex phenomenon associated with a rapid rise in gallbladder pressure.
似乎有与胆囊压力急剧升高有关的反射现象。

4. It 引导的句子。科技英语中，可能会出现形式主语 it 引出的句子，基本结构为 it+be+adj. /v. -ed + to do sth. /that clause。其中，it 是形式上的主语，真正的主语是后面的 to do sth. 或者 that clause。翻译成汉语时，可以按汉语的习惯，先给出话题，然后再进行评价。这种汉语意义的"主题+评价"关系，并非英语中的主谓关系，所以属于汉语无主句。

In cases where it is not economical to remove the surface, shafts are dug into the earth, with side tunnels to follow the layer of ore.

如果露天开采不经济，则顺着矿层，朝地下挖掘矿井，开凿侧向巷道。

It has actually become more economical in some cases to ship steel across the ocean than to produce it in older U. S. plants.

实际上在某些情况下，从海外跨洋将钢铁运到美国比在美国的老旧钢厂里生产钢铁还要更经济。

上述两例中，原文用 it 作形式主语，紧接着是"连系动词+表语"构成的评价性谓语，然后引出真正的主语，是典型的西方形合思维模式。译句先给出话题，然后对话题进行评价，是典型的汉语意合思维模式，译文更直接、简练。

◎ 练习五　将下面的句子翻译成汉语。

1. There has been reported that the intestinal microecological balance is broken in COVID-19 patients, manifesting a significant reduction of the intestinal probiotics such as lactobacillus and bifidobacterium.

2. Blast furnace ironmaking consumes most of the energy in an iron and steel process; therefore, it is important to further reduce the consumption of the reducing agents.

3. It is worth noting that the Young's modulus or shear modulus can be reduced to improve calculation efficiency only when it has a little influence on the results.

4. Owing to the irregular shapes of sinter and coke, it is difficult to measure and calibrate the contact parameters.

5. Only in this way can we obtain more accurate burden flow, segregation, and distribution data under different burden structures.

6. Xu et al. established a 4,070 m^3 blast furnace model using the DEM, and analyzed the volume percentages of the sinter, pellet, and lump ore in the radial, circumferential, and vertical directions of the hopper.

7. They discovered that when filling the hopper, the small particles were prone to distribution in the center, and moved less to the edge of the hopper. The distribution of large particles presented an opposite trend to that of small particles; no segregation occurred for intermediate particles.

8. The reduction of heat losses is achieved through the thermal insulation of the tuyere surface, and the increase in the completeness of the fuel combustion is achieved with the improvement of the mixture homogeneity after mixing fuel with hot blast.

9. Known methods for creating a heat-shielding layer on the surface of BF tuyeres include the plasma spraying of a ceramic layer on the nose part's outer surface. However, owing to the lack of sustainable adhesion of the tuyere material to the sprayed layer, the coating exfoliated, and hence this method was not widely used.

10. Although researchers have conducted detailed studies on the flow and segregation of the

burden; owing to the limitations of the computing power, model algorithms, and calibration of contact parameters, the existing research on the flow and segregation of the burden can be further improved.

11. To increase the efficiency of mixing NG with the blast, it is advisable to supply it closer to the axis of the air passage and in several places using various types of swirlers. Taking into account the simultaneous temperature increase in the air passage, the thermal insulation of the inner nozzle is required.

12. When the burden moves in the chute, the small particles are mainly distributed at the bottom of the chute, the large particles are distributed above the small particles, and the angular velocity of the larger particles is greater at the end of the chute. There are more large particles on the outer side of the burden trajectory, and more small particles on the inner side; the angular velocity of large particles is larger, and they roll more easily.

第六课　直接还原工艺

> 一个粮食，一个钢铁，有了这两个东西就什么都好办了。
>
> ——毛泽东①

一、科技术语

direct reduction Processes for the production of iron and steel directly from ore by using reducing gases without the smelting of pig iron in blast furnaces.（直接还原；直接还原法）

sponge iron Also called direct-reduced iron（DRI）, produced from direct reduction of iron ore（in the form of lumps, pellets or fines）by a reducing gas produced from natural gas or coal. The reducing gas is a mixture majority of hydrogen and carbon monoxide which acts as reducing agent.（海绵铁）

wrought iron Commercially pure iron, a tough, malleable, ductile, easily-welded and fibrous material due to the slag inclusions（a normal constituent）. In contrast to steel, it has very low carbon content.（熟铁；锻铁）

powder metallurgy The process of blending fine powdered materials, pressing them into a desired shape or form（compacting）, and then heating the compressed material in a controlled atmosphere to bond the material（sintering）. The powder metallurgy process generally consists of four basic steps：powder manufacture, powder blending, compacting, and sintering.（粉末冶金；粉末冶金学；粉末冶金技术）

thermal decomposition The breaking down of a chemical compound by heat into smaller components which do not recombine on cooling.（热分解）

reducing agent Also called a reductant or reducer, an element or compound that loses（or "donates"）an electron to another chemical species in a redox chemical reaction. Since the reducing agent is losing electrons, it is said to have been oxidize.（还原剂）

---

① 出自毛泽东在1958年8月13日视察天津时的谈话。

pot furnace A furnace in which the charge is contained in a pot or crucible.(坩埚炉)

reverberatory furnace An industrial smelting furnace in which heat is transferred to the material by radiation from the gaseous products of fuel combustion, as well as from the incandescent interior surface of the refractory furnace lining.(反射炉)

regenerative furnace A furnace in which the incoming air is heated by regenerators.(蓄热氏炉)

shaft furnace A vertical, refractory-lined cylinder in which a fixed bed (or descending column) of solids is maintained and through which an ascending stream of hot gas is forced; for example, the pig-iron blast furnace and the phosphors-from-phosphate-rock furnace.(竖炉)

rotary kiln A pyroprocessing device used to raise materials to a high temperature (calcination) in a continuous process. Materials produced using rotary kilns include: cement, lime, refractories, metakaolin, titanium dioxide, alumina, vermiculite, iron ore pellets.(回转窑)

retort furnace A closed laboratory vessel with an outlet tube, used for distillation, sublimation, or decomposition by heat.(甑式炉)

electric furnace Any furnace in which the heat is provided by an electric current.(电炉)

fluidized-bed reactor A type of reactor device that can be used to carry out a variety of multiphase chemical reactions. In this type of reactor, a fluid (gas or liquid) is passed through a granular solid material (usually a catalyst possibly shaped as tiny spheres) at high enough velocities to suspend the solid and cause it to behave as though it were a fluid. This process, known as fluidization, imparts many important advantages to the FBR. As a result, the fluidized bed reactor is now used in many industrial applications.(流化床反应器)

二、英语原文

Direct Reduction Processes

1 Due to the desire (and in some countries the economic necessity) to employ the lower grade ores and available fuels (non-coking bituminous coals, anthracite, lignite, etc. that are unsuitable for blast furnace fuels), an intensive search has been underway in many quarters to develop a process or processes that would at least partially supplant the blast furnace as a source of iron for steelmaking. Such processes are referred generically as direct reduction process and their purposes are to either: (1) produce steel directly from iron ore; (2) make a product

equivalent to blast furnace pig iron for use in present steelmaking processes; or (3) produce low carbon iron as a melting stock (sometimes referred to as synthetic scrap) for making steel by existing processes. In spite of the intensity of effort, no commercially useful direct reduction process has been developed to date that shows any immediate promise of supplanting the blast furnace as the chief source of iron units for steelmaking on a large scale.

Principles of Direct Reduction

2 Dry, pure iron oxides (Fe_2O_3 or Fe_3O_4) in the presence of reducing substance, form dark-gray, porous masses having the same size and shape as the original lumps of particles when reduced at temperatures below 1 650 °F (900 °C). A temperature of 1,740 °F to 1,830 °F (950 °C to 1,000 °C) is necessary to effect complete reduction in a reasonable length of time, unless the ore is of selected particle size.

3 At 1 830 °F (1,000 °C) the product begins to sinter. At 2 190 °F (1,200 °C), a pasty, porous mass forms. At 2,370 °F (1,300 °C) the mass absorbs carbon rapidly, if the latter is present, and begins to fuse. The upper melting (entirely liquid) point of pure iron is 2,785 °F (1,530 °C).

Sponge Iron

4 Sponge iron provided the main source of iron and steel many centuries before the blast furnace was developed around 1300 A.D. Sponge iron was produced in relatively shallow hearths or in shaft furnace, both of which used charcoal as fuel.

5 The product of all of these smelting processes was a spongy mass of coalesced granules of nearly pure iron intermixed with considerable slag. Usable articles of wrought iron were produced by first hammering the spongy mass while still hot from the smelting operation, to expel most of the slag and compact the mass. By repeated heating and hammering, the iron was further freed of slag and forged into the desired shape. All of the methods whereby low carbon wrought iron could be produced directly from the ore were called direct processes.

6 After development of the blast furnace, which made large quantities of iron having a high carbon content available, low carbon wrought iron was produced by refining this high carbon materials; because two or more steps were involved in the processes employed, they came to be known as indirect processes.

7 Direct methods are still in use and have never been wholly abandoned even by the most advanced nations. The ease with which iron ores are reduced makes the direct process appear enticingly simple and logical. The reduction takes place at low temperature and absorbs little

heat, some of the reactions actually being exothermic.

8   In modern times sponge iron has found increasing use in various industrial processes other than in the manufacture of wrought iron. The iron produced in sponge form has a very high surface area and is used in the chemical industry as a strong reducing agent. It is chemically much more active than steel or iron in the form of millings, borings, turnings or wire.

9   Sponge iron may be produced as a granular material or as a sintered mass, depending upon the methods of manufacture. In the purified granular form, commonly known as powdered iron, it is used in the manufacture of many useful articles by the techniques of powder metallurgy.

10   The iron powders are compacted by pressure into the approximate shape of the finished article, then sintered at 950-1,095℃ (1,740-2,000℉) in furnaces provided with a protective atmosphere to prevent oxidation and finally pressed or machined to their final shape. Iron powders are produced not only by direct reduction of iron ores or oxides using solid carbonaceous reducing agents and gaseous reducing agents such as carbon monoxide and hydrogen, but also by electrolytic processes and by thermal decomposition of iron carbonyl, $Fe(CO)_2$.

11   The production of sponge iron demonstrates that iron ore can be reduced directly with relative ease on a small scale. While the process of producing low carbon iron directly from ore is theoretically attractive and appears more logical than indirect processes, direct processes have so far failed in competition with indirect methods.

12   Apparently, the chief reasons are: (1) The ore must be very rich. (2) The ore must be finely divided and intimately mixed with or carefully placed over the reducing agent. (3) So far no practical plan has been evolved in which the ore and reducing agent may be mixed in proper proportions to leave no excess of either. If an excess of the former is permitted, the process is wasteful of ore, while in the presence of an excess of the latter, the iron is obtained at a low temperature in sponge or pasty form which is hard to handle. If produced at a higher temperature phosphorus in the ore will be reduced and the carbon absorbed from the fuel, give a very impure metal, little better than pig iron.

13   The attempts to develop large scale direct processes have embraced practically every known type of apparatus suitable for the purpose including pot furnaces, reverberatory furnaces, regenerative furnaces, shaft furnaces, rotary and stationary kilns, retort furnaces, rotary hearth furnaces, electric furnaces, various combination furnaces, fluidized bed reactors and plasma reactors.

14   Many reducing agents including coal, coke, graphite, char, distillation residues, fuel oil, tar, producer gas, coal gas, water gas and hydrogen have also been tried.

15 In considering the direct-reducing processes now in use, a proper perspective is obtained if it is remembered that the ferrous products presently made by direct reduction represent less than 2 percent of the total iron and steel production of the world.

16 No effort will be made here to evaluate or compare the different processes from either the economic or technical standpoints. Those that were developed to meet purely local conditions with some degree of success might prove to be completely impractical under other conditions involving, for example, different fuels and different raw materials.

三、分析与讲解

直接还原工艺(Direct Reduction Iron, DRI)指在低于矿石熔化温度下，通过固态还原，把铁矿石冶炼成铁的过程。通过这种工艺生产的铁保留了失去氧时形成的大量微气孔，在显微镜下观看形似海绵，因此也被称为海绵铁。用球团状矿制成的海绵铁也称为金属化球团。直接还原铁中能氧化发热的元素如碳、硅、锰的含量很少，不能用于转炉炼钢，但适用于电弧炉炼钢，因此就形成一个"直接还原炉——电炉"的钢铁生产新流程。

我国对直接还原铁工艺的探索、开发、研究已超过50年，但因铁矿、煤炭、气源等原燃料条件的限制，我国直接还原铁工艺发展的实际成效甚微。不过国内铁矿精选技术、煤制气技术日益成熟，为直接还原铁工艺在我国的发展奠定了可靠的资源和技术基础。

直接还原铁可完美替代废钢，有利于调整钢铁产品结构，提高产品质量，符合我国钢铁行业节能减排政策①，这一工艺值得大力发展。以下是对文章的分析与讲解：

第1段：(1)第1句中的 desire to employ，employ 好翻译，desire 对应的汉语表达有"渴望""愿望"等，但"渴望"情感意义太强，"愿望"一般作名词用，因此这里最好译成"希望"或者"需要"。后面的 available 作定语时一般后置，但这里前置，就表示"拥有的"。若将 the lower grade ores 译成"低品位矿石"，则最好将 available fuels 译成"手头拥有的燃料"或"自身拥有的燃料"，这样两个前置定语有个对应，避免一个较长而另一个较短。(2) a process or processes 如果直译成"一种或多种工艺"，显得生硬。直接译成"工艺"，将原文名词的单复数概念模糊化，更符合汉语的表达习惯。(3)第2句中的 either... or... 一般译为"或者是……或者是……"。但这里涉及三个选项，either 和 or 离得太远，如果依然这样翻译，两个"或者是"离得太远，所以不妨译成"可能是"，不仅再现了原文内涵，又简洁易懂。(4)翻译 intensity of effort, immediate promise, commercially useful 时，最好采用词性转换的翻译技巧，将它们分别译为"力度很大""有希望立刻"和"具有商业用途"。

---

① 宋赞，李相帅，查春和. 我国直接还原铁工艺的发展现状及趋势[J]. 冶金管理，2020 (8)：22-23.

第 2 段：(1)第 1 句中 in the presence of reducing substance 虽然字面意思是"在有还原剂存在的情况下"，但实际意思是"还原剂发挥了还原的作用"，所以应译为"在还原剂的作用下"。后面的 Fe_2O_3、Fe_3O_4、900℃等都是国际通行的表达形式，因此采用零翻译的翻译策略，直接在括号里保留原文。(2)第 2 句中 complete reduction 当然指前面所说的 Dry，pure iron oxides，所以翻译时需要增加主语"铁氧化物全部还原"。后面的 a reasonable length of time 不能机械地翻译，而要灵活翻译为"较长的时间"。

第 3 段：(1)第 3 句中，if the latter is present 意思是"如果存在碳的话"，因为是补充说明，所以最好在其之前增加破折号，这是中文补充表达惯用的手段。(2)第 4 句中 entirely liquid 字面意思是"完全液体"，也就是"全部熔化"。

第 4 段：第 2 句中，relatively shallow 只修饰 hearths，如果将 in relatively shallow hearths or in shaft furnace 译为"在较浅的炉床或竖炉中"，则容易引起误解。为避免歧义，"竖炉"前也需要使用"在"。

第 5 段：(1)查阅英汉词典，coalesce 的释义有"合并；联合；接合"，但这三个表达法都不适合用在译句中。又查英英词典，该词有以下意思：① To come or grow together into a single mass（E. g. the material that coalesced to form stars）；② To come together as a recognizable whole or entity（E. g. the stories that coalesced as the history of the movement）；③ To come together for a single purpose（E. g. The rebel units coalesced into one army to fight the invaders.）。结合上下文，原句中 coalesce 应理解为"to come or grow together into a single mass"，应译为"组成"。(2)第 2 句虽然是个简单句，但结构比较复杂，应该先译方式状语 by first hammering the spongy mass while still hot from the smelting operation，然后翻译目的状语 to expel most of the slag and compact the mass，最后再译主句 Usable articles of wrought iron were produced，这样层次分明，有条不紊。

第 6 段：which made large quantities of iron having a high carbon content available 是非限制性定语从句，是后面主句中 low carbon wrought iron was produced by refining this high carbon materials 能够实现的先决条件，所以应先翻译该定语从句："高炉的研制使得大量生产高碳铁成为可能"。

第 7 段：(1)第 2 句中的 ease，可以直译成"方便地"，也可以正说反译，译成"不费劲地"；enticingly simple 字面意思是"诱人得简单"，也就是"简单诱人"。(2)第 3 句中，exothermic 意思是"发热的，放出热量的"。这个词是由前缀 exo-加 thermic 构成的，exo-表示"outside, external"等意思，如：exocentric(离心的)，exocarp(外果皮)，exoatmosphere (外大气层)，exocyclic(环外的)，exodus(出去，外出)，exotic(外国的，外来的)，exodermis(外皮层)。

从事科技翻译的人士应熟悉科技词汇中常用的前缀、后缀等，以便快速提高科技词汇量。常见的前缀有：

A

a-(无，非，不)：acentric 无中心的；atypical 非典型的；aperiodic 非周期的、不定期的

ab-(离开，相反，不)：abaxial 离开轴心的；abnormal 异常的

amphi-(双，两)：amphibian 两栖的；amphibiology 两栖生物学；amphidiploids 双二倍体(具有两套染色体的生物)

an-(无，不)：anelectric 不导电的；anelectrolyte 非电解质；anoxia 缺氧症

ante-(前，先)：antecedent 前项；antechamber 预燃室；antevert 前倾

anti-(反对，相反，防止)：antiaging 防衰老；anticlockwise 逆时针的；antimissile 反导弹的

auto-(自己，自动)：antojigger 自耦变压器；automatic 自动的；autometer 汽车里程表

B

bi-(两，二)：binary 二进制的；binuclear 双核的；birefraction 双折射

bio-(生物，生命，生活)：biochronometer 生物钟；biocide 杀虫剂；biomolecule 生命分子

by-(次要的，附带的，副的)：by-channel 支架；by-effect 副作用；bypath 侧管

C

circum-(周围，环绕)：circumference 圆周，四周；circumlunar 环月的；circumterrestrial 绕地球的

co-(共同)：cochannel 同频道的；cocurrent 平行电流；coenergy 同能量

counter-(反对，反抗，逆，重复，副)：counteractant 中和剂；counter-air 防控的；counter-arch 反拱；counter-fire 逆火；couterstain 复染；counter-shaft 副轴

D

de-(否定，非，脱)：decoder 解码器；decoloration 脱色；dehydration 脱水；decomposition 分解

deca-(十)：decameter 十米；decaploid 十倍体；decatrack 十轨系统

deci-(十分之一)：decigram 分克；decimal 十进制的；decimillimeter 丝米

di-(二，双)：diatomic 双原子；dioxide 二氧化物；dicarbide 二碳化物

dia-(横过，通，全)：diagonal 对角线；diameter 直径；diathermal 透热的

dis-(不，无，相反)：disabled 报废的；discharger 排气装置；disconformity 不一致、不相称

dys-(恶化，不良，困难，障碍，疼痛)：dysbiosis 生态失调；dysbolisn 代谢障碍；dysomnia 睡眠困难；dysaethesia 触物感痛

E

ef-(出，离去)：efflation 吹出；effluence 流出；effoliation 落叶
endo-(内)：endogen 内生植物；endolymph 内淋巴；endoparasite 体内寄生虫
ennea-(九)：enneagon 九角形；enneahedron 九面体；enneode 九极管
ex-(出，外)：encavate 发掘；exclude 排外；extract 抽出
exo-(外，外部)：exobiology 外太空生物学；exosphere 外大气层；exoskeleton 外骨骼
extra-(以外，超过)：extrados 外拱线；extra-low 超低的；extraneous 外部的

H

hecto-(百)：hectoampere 一百安培；hectogram 一百克；hectowatt 一百瓦
hemi-(半)：hemicycle 半圆形；hemiparasite 半寄生物；hemisphere 半球
hepta-(七)：heptagon 七角形；heptahedron 七面体；hepatode 七级管
hetero-(异)：heteromorphic 异形的；heteropolar 异极的；heterosexual 异性的；
hexa-(六)：hexagon 六角形；hexangular 有六角的；hexode 六极管
holo-(全)：hologram 全息图；holohedron 全面体；holophote 全反射镜
homo-(同，恒)：homocentric 同中心的；homothermic 恒温的；homotype 同型
hyper-(超过，过多)：hyperacid 酸过多的；hyperbar 超高压；hyperellipsoid 超椭圆体
hypo-(下，次，低，少)：hypochromatic 染浅色的；hypoelastic 次弹性的；hypogravity 低重力

I

infra-(下，低)：infrared 红外线；infrasonic 低于声频的；infrastructure 基础设施
inter-(相互)：intercontinental 洲际的；intermix 混杂的；interweave 混纺
iso-(等，同)：isoelectric 等电子的；isogon 等角多边形；isotope 同位素

M

macro-(大，宏，长)：macroclimate 大气候；macrophysics 宏观物理学；macropod 长尾的

mal-(恶，不良，不)：malformation 畸形；malfunction 功能失调；malnutrition 营养不良

micro-(微)：microbiology 微生物学；microscope 显微镜；microwave 微波

milli-(千分之一，毫，千)：milligram 毫克；millimeter 毫米；millennial 一千年的

mini-(小)：minimum 最小值；minisub 小型潜艇；miniwatt 小功率

mono-(单一，独)：monatomic 单原子的；monocolor 单色；monoxide 一氧化物

multi-(多)：multipurpose 多种用途的；multi-roll 多辊；multishaft 多轴

O

omni-(全，总，都)：omnidirectional 全向的；omnifactor 多因素；ominiforce 全向力

P

paleo-(古，旧)：paleoanthropology 古人类学；paleobotany 古植物学；paleozoology 古动物学

penta-(五)：pentagon 五角形；pentode 五极管；pentoxide 五氧化物

peri-(周围，外层，靠近)：pericentral 中心周围的；periderm 外皮；perihelion 近日点

poly-(多)：polyatomic 多原子的；polycrystal 多结晶体；polygon 多角形

proto-(原始)：protoplanet 原行星；protosoil 原生土；prototype 原型

pseudo-(假)：pseudo-atom 膺原子；pseudo-earthquake 假地震；pseudo-effect 伪效应

Q

quasi-(类似，准，半)：quasi-conductor 半导体；quasi-norm 拟范数；quasi-saturation 准饱和

S

semi-(半)：semiautomatic 半自动的；semicircle 半圆；semiconductor 半导体

sept-(七)：septangle 七角形；septfoil 七叶形；septuple 七倍的

sex-(六)：sexangle 六角形；sexfoil 六叶形；sexvalence 六价

stereo-(立体)：stereophony 立体音响；stereosonic 立体声的；steogram 立体图

sub-(下，次，低)：submarine 潜水艇；subsonic 亚音速的；subcarbide 低碳化物

supra-(超，上)：supraconductivity 超导电性；suprafluid 超流体的；supramolecular 超分子的

syn-（共同，相同）：synchronism 同步性；synclastic 同方向的；sync-pulse 同步脉冲

T

tetra-（四）：tetragon 四角形；tetrode 四极管；tetroxide 四氧化物
tri-（三）：triangle 三角形；tricolor 三色的；trisection 三等分

U

ultra-（极端，超，以外）：ultrabandwidth 频带特别宽的；ultrasound 超声波；ultraviolet 紫外线

学习科技翻译的人士不妨自己收集、整理构成科技词汇的常见后缀并花点时间记牢，这对于理解科技英语、从事科技翻译工作非常有益。

第 8 段：(1) 第 2 句中的 the iron produced in sponge form 其实就是 sponge iron，这样表达是为了让语言形式更丰富。一般来说，科技英语与文学英语不同，不太追求表达形式的多样性，尤其是在术语的使用上，更看重一致性（consistency），即术语表达形式应从头至尾保持一致，以免读者误以为作者在讨论不同的内容。但偶尔这样表达也未尝不可，只要不造成读者理解上的混乱。(2) millings（铣屑）、borings（钻屑）、turnings（车屑）等常作为普通词汇使用，但在这里是专业术语，不可误将其当作普通词汇翻译。

第 9 段：(1) 汉语中一般不说"生产方法"，所以第 1 句中的 methods of manufacture 最好译为"生产工艺"。翻译时，要始终注意英语和汉语词语在搭配上的差异。(2) 第 2 句事实上是先给 powdered iron 下定义，然后再说明 powdered iron 的功用。翻译时，也应先给 powdered iron 下定义："海绵铁提纯后若呈粒状形式，通常称为铁粉。"

第 10 段：第 2 句中第一个 by 后面的介词宾语较长，为避免译句过长，读来难以消化，可将 such as carbon monoxide and hydrogen 的译语放在括号中。英语原句用三个 by 引出三种铁粉生产工艺，其中，第二个 by 短语和第三个 by 短语用 and 连接，但在汉语中，当三种工艺中任一种都可以生产出铁粉时，一般不用"和"，而用"或"，所以第三个 by 短语前的 and 应译成"或"。

第 11 段：第 2 句中的 while 表示的不是时间，而是让步，所以应译成"尽管"。后面的 failed，不能译为"失败"，更好的译法是"处于下风"。

第 12 段：(1) The ore must be very rich 后面明显省略了 in iron，所以需译为"矿石含铁量必须很高"。(2) 翻译 If an excess... which is hard to handle，应力求简洁。为此，可将 If an excess of the former is permitted 译成汉语无主句"如果允许铁矿石过剩"，而将 While in the presence of an excess of the latter 译成条件句"如果还原剂过剩"或"如果允许还原剂过

剩"。(3)因为前面已经有了两个条件句,那么不妨将 If produced at a higher temperature 译为形式类似的条件句("如果提高温度"),使上下文衔接紧密,更具可读性。

第13段:只有一句话,虽然比较长,但结构比较简单。翻译时,最好将 practically 译在句首,将 attempts 译成动词:"事实上,通过直接法大规模炼铁曾尝试使用过每一种可用于该目的的设备……"将句中副词提取出来译在句首,也是一种常用的翻译技巧。

第15段:considering 字面意思是"考虑",但这里隐含的意思是"评价""判断",所以可译为"对……作判断"。A proper perspective 字面意思是"正确的视角",这里应理解为"正确的理解"。

第16段:第1句有两种译法:一是按照原文的句序翻译"这里将不对各种生产工艺做评价或比较,无论是从经济还是从技术的角度";二是改变原句语序,译成"这里将不从经济或技术的角度对各种生产工艺做评价或比较"。虽然第二种译法更简洁,但第一种译法更贴切,因为第二种译法似乎隐含了只有经济和技术两种比较视角。

四、参考译文

直接还原工艺

1　因为希望(在某些国家,是经济上的需要)利用低品位矿石和手头拥有的燃料(无法炼焦、不适宜做高炉燃料的烟煤、无烟煤、褐煤等),世界上很多地方都开展了深入的研究,(试图)研发出至少可以部分取代高炉为炼钢提供铁原料的工艺。这些工艺总称为直接还原工艺,其目的可能是:(1)直接从铁矿石生产钢;(2)生产出与高炉生铁相同的产品用于现有的炼钢工艺;(3)生产出低碳铁,用作现有炼钢工艺的熔料(有时被称为人造废钢)。尽管研究的力度很大,但至今也未研制出具有商业用途、有希望立刻取代高炉成为大规模炼钢生产主要铁源的直接还原工艺。

直接还原的原理

2　在还原剂的作用下,干燥的纯铁氧化物(Fe_2O_3 或 Fe_3O_4)在低于1650℉(900℃)时开始被还原,生成深灰色、多孔的块状物质,大小和形状与原颗粒块相同。铁氧化物全部还原,则需要温度达到1740℉至1830℉(950℃至1000℃),且需要较长的时间,除非铁矿石的粒度经过了严格挑选。

3　1830℉(1000℃)时,产品开始烧结。2190℉(1200℃)时,一种糊状的、多孔的物质形成。2370℉(1300℃)时,形成的物质迅速吸收碳——如果存在碳的话——并开始熔化。纯铁的上熔点(全部熔化)是2785℉(1530℃)。

海绵铁

4 在高炉于公元 1300 年左右被研制出来之前，海绵铁一直是多个世纪里钢铁的主要来源。当时，海绵铁是在较浅的炉床或在竖炉中生产的，两者都使用木炭做燃料。

5 这些冶炼工艺得到的产品是一种由海绵状的、接近纯铁的微粒组成的块，夹杂有相当多的渣成分。可以通过捶打刚从冶炼炉出来的灼热的海绵块，除去大部分渣成分，使海绵块更致密，从而制造有用的熟铁物件。通过反复加热和捶打，海绵铁中的渣成分更少，海绵铁也被锻造成所需的形状。所有直接从矿石生产低碳熟铁的工艺都被称为"直接工艺"。

6 高炉的研制使得大量生产高碳铁成为可能。此后，低碳熟铁的生产便通过精炼这种高碳铁来实现。因为这类工艺包含两个甚至更多的步骤，所以被称为"间接工艺"。

7 直接法今天仍在使用，甚至最发达的国家也没有完全抛弃直接法。直接法便于还原铁矿石，显得简单诱人，因此使用直接法就显得顺理成章了。还原是在低温下发生的，吸收热量很少，某些反应事实上还放热。

8 今天，除用于制造熟铁外，海绵铁在各种工业流程中的应用日益广泛。海绵铁表面积较大，它作为强还原剂被用于化学工业。海绵铁的化学性质比以磨粉、钻粉、削屑或金属丝形式存在的钢和铁活泼得多。

9 海绵铁的形态取决于生产工艺，生产出来后可能是粒状，也可能是烧结成块。海绵铁提纯后如呈粒状形式，通常被称为铁粉。借助粉末冶金技术，铁粉可制成很多有用的物品。

10 将铁粉压紧，压成接近成品的形状，然后在 950℃ 至 1095℃（1740°F 至 2000°F）的炉中，在保护性气氛下烧结，以防氧化，最后压制、加工成最终形状。生产铁粉不仅可以采用固体含碳还原剂和气体还原剂（如一氧化碳和氢）直接还原铁矿石或铁氧化物，还可以采用电解工艺或通过羰基铁（$Fe(CO)_2$）热分解方法。

11 海绵铁的生产表明，铁矿石可以相对方便地小规模直接还原。尽管直接从矿石中生产低碳铁理论上吸引人，听起来也比间接法符合逻辑，但迄今为止，直接法在与间接法的竞争中一直处于下风。

12 很明显，主要原因有：(1) 矿石含铁量必须很高。(2) 矿石必须加工成很细的粉末，并且要与还原剂很好地混合，或者小心地放在还原剂之上。(3) 迄今为止，还没有人提出实用的方案，确保矿石和还原剂以适当比例混合，以避免两者中任何一种无过剩。如果允许铁矿石过剩，则浪费矿石；如果还原剂过剩，那么低温下得到的铁会呈海绵或糊状，难以处理。如果提高温度，矿石中的磷将被还原，燃料中的碳将被吸收，生成不纯净的金属，其质量并不比生铁好。

13 事实上，通过直接法大规模炼铁曾尝试使用过每一种可用于该目的的设备，包括坩埚炉、反射炉、蓄热氏炉、竖炉、回转窑和固定窑、甑式炉、转底炉、电炉、各种组合炉、流化床反应器和离子反应器。

14　也尝试使用过各种各样的还原剂，包括煤、焦炭、石墨、木炭、蒸馏残余物、燃料油、焦油、发生炉煤气、煤气、水煤气和氢。

15　在对当前应用的直接还原工艺作判断时，如果记得直接还原生产的铁制品只占世界钢铁总产量的不到2%，就会有一个正确的理解。

16　这里将不对各种生产工艺作评价或比较，无论是从经济还是从技术的角度。那些纯粹为了满足当地生产条件而开发的工艺在当地获得了某种成功，但在不同的条件下，比如使用不同的燃料和原材料时，该生产工艺可能就会完全不实用。

五、专题讨论

科技英语中定语从句的翻译

定语从句是英语最突出的语法现象之一，定语从句的大量使用是英语区别于汉语的主要句法特征之一。①

定语从句就是用一个句子来做定语。定语从句中同样有主语、谓语，有时还有宾语及其他成分。为了将主句和从句区分开来，有时也是为了表达的需要，定语从句会有引导词（关系代词that，which，who，whose，whom，as和关系副词when，where，why）来引导，这些引导词反映了定语从句与先行词之间的关系。

一方面，科技英语继承了科学活动所具有的理性思维，强调表述清晰、言简意赅，在言说过程中尽量消除人文色彩与修辞手法，具有客观性的特点。② 另一方面，科技英语用于表达科学原理、规律、概念以及各事物之间错综复杂的关系，讲究逻辑严密，无懈可击，所以结构复杂的长句使用得较多，③ 尤其是带有定语从句的长句。此外，由于英语是句尾开放，定语从句本身还可以带有自己的定语从句，在语感允许的范围内无限延伸，这就使得定语从句的翻译常常不易。因此，简洁、清晰地翻译科技英语，离不开对定语从句

---

① 汉语中也有定语从句，如"他卖给我的旧手机是华为产的"，这个句子中，"他卖给我的"就是句子作定语。当然也有学者认为汉语中不存在定语从句，如包彩霞在"汉语的定语与英语的定语从句"（《北京第二外国语学院学报》2004年第2期）中认同说"而汉语中则不存在定语从句"。如果这样认为，"他卖给我的"算是由主谓成分构成的词组。我们倾向于前一个观点。即便如此，汉语中定语从句的使用仍远不如英语中普遍。

② 吕英莉. 科技英语的翻译特点与翻译策略研究[J]. 有色金属工程，2022(7)：195-196.

③ 蒲筱梅. 科技英语定语从句汉译的句法重构[J]. 云南农业大学学报，2010(8)：91.

的有效处理。例如：

Due to the desire (and in some countries the economic necessity) to employ the lower grade ores and available fuels (non-coking bituminous coals, anthracite, lignite, etc. <u>that are unsuitable for blast furnace fuels</u>), an intensive search has been underway in many quarters to develop a process or processes <u>that would at least partially supplant the blast furnace as a source of iron for steelmaking</u>.

因为希望(在某些国家是经济上需要)利用低品位矿石和自身拥有的燃料(无法炼焦、<u>不适宜做高炉燃料的</u>烟煤、无烟煤、褐煤等)，世界上很多地方都开展了深入的研究，以研发出<u>至少可以部分取代高炉为炼钢提供铁原料的</u>工艺。

这个例子中，原文画线部分为两个定语从句，分别修饰前面的名词短语。两个定语从句的使用使整个句子变长，结构也变得复杂。译句中的画线部分是定语从句对应的译文。可以看出，如何处理好这两个定语从句是翻译整个句子的关键。翻译时，它们都被译为前置修饰语，放在被修饰对象前，意思表达清楚，也与汉语定语的使用习惯保持一致。译文也层次清晰，通顺达意。

总的说来，英语定语从句的翻译主要有四种方法：

1. 译为前置定语。将英语定语从句译成汉语中心词的前置定语。例如：

During the oxidation period, the reactions <u>that occur in the bath of the basic electric-arc furnace</u> are similar to those in the basic oxygen furnace.

氧化期中，<u>碱性电弧炉熔炉中发生的</u>反应与碱性氧气转炉中的反应相似。

原句中，定语从句 that occur in the bath of the basic electric-arc furnace 作为后置定语修饰先行词 the reactions，被译为"碱性电弧炉熔炉中发生的"放到"反应"之前，作"反应"的前置定语。

The higher the carbon content, the higher strength and the greater the hardness <u>to which the steel may be heat-treated</u>.

含碳量越高，抗拉强度就越高，<u>热处理后钢所达到的</u>硬度也越高。

定语从句 to which the steel may be heat-treated 修饰先行词 the hardness，被译为"热处理后钢所达到的"放到"硬度"之前，作"硬度"的前置定语。

Processes which produce a molten product similar to blast-furnace hot metal directly from the ore are defined as direct-smelting processes.

直接从矿石中生产类似于高炉铁水的熔融产品的工艺，被称为直接熔炼工艺。

定语从句 which produce a molten product similar to blast-furnace hot metal directly from the ore 被译为"直接从矿石中生产类似于高炉铁水的熔融产品的"放在"工艺"之前，成为"工艺"的前置定语。

2. 译为后置独立句子或小句。基本按照原语序翻译原句，将英语定语从句翻译成独立的句子或小句，放在原主句之后。例如：

Production of aluminum from bauxite involves two distinct processes which are often operated at quite different locations.

从铝土矿中提取铝包括两个截然不同的过程，这两个过程在完全不同的地点进行。

定语从句 which are often operated at quite different locations 被译为独立句子，放置在原主句之后，重复先行词"这两个过程"作为分句的主语，句意完整、清楚。

Circulating scrap consists of sheared ends and rejected materials which are normally returned immediately to the steelmaking vessel.

循环废钢包括各种切头和废品，它们通常会很快回到炼钢炉中。

翻译此句时，假如把定语从句"通常会很快回到炼钢炉中"放到先行词"各种切头和废品"之前，会导致定语太臃肿，不利于清晰表达。按照原句语序进行翻译，将定语从句译成并列分句，放于原主句之后，由"它们"充当分句的主语，这样意思清楚，衔接自然。

Direct reduced iron derived from virgin iron ore units is a relatively pure material which dilutes contaminants in the scrap and improves the steel quality.

由天然铁矿石冶炼而成的直接还原铁是一种相对纯净的材料，可以稀释废钢中的杂质，提高钢的质量。

定语从句 which dilutes contaminants in the scrap and improves the steel quality 太长，被译为独立小句依然放在先行词之后，进一步说明其特性，句间衔接自然。

3. 融合法。把定语从句和它所修饰的先行词融合在一起翻译。例如：

After development of the blast furnace, which made large quantities of iron having a high carbon content available, low carbon wrought iron was produced by refining this high carbon materials.

高炉的研制使得大量生产高碳铁成为可能。此后，低碳熟铁的生产便通过精炼高碳铁而实现。

将非限定性定语从句 which made large quantities of iron having a high carbon content available 与先行词 development of the blast furnace 融合在一起翻译为"高炉的研制使得大量生产高碳铁成为可能"，独立成句。同时，用"此后"译出 after 的意思，以明确前后句子之间的逻辑关系，使译文自然、达意。

After works experiments which started in the 1940s, Imperial Smelting Corporation, at Avonmouth in England announced in 1957 the successful development of a blast furnace for making zinc.

英国埃文茅斯的帝国熔炼公司从 20 世纪 40 年代开始，经过多次工厂试验，于 1957 年宣布成功研发了鼓风炉炼锌。

定语从句 which started in the 1940s 与先行词 works experiments 及之前的介词 after 一起合译为"从 20 世纪 40 年代开始，经过多次工厂试验"，译文流畅，符合汉语表达习惯。

4. 译为状语。英语定语从句意义非常广泛，虽说在结构上作定语，但还可以表示"原因""条件""假设""让步""目的""结果""时间""转折"等意义。定语从句通常需要译为状语，来表示原因、结果、目的、条件等。例如：

It is extremely attractive to develop iron and steel-making plant which is more flexible in operation than the blast furnace and which can therefore react more quickly to market demands or to problems of supply.

因为采用比高炉操作更灵活的冶炼技术，可以对市场需求或供应问题作出更快的反应，所以建设这样的钢铁厂极具吸引力。

定语从句 which is more flexible in operation than the blast furnace and which can therefore react more quickly to market demands or to problems of supply 太长，暗示主句 It is extremely attractive to develop iron and steel-making plant 的原因，译为原因状语从句，放在主句之前，逻辑清楚，表达自然。

There may be the possibility of using very high grade iron ore, which can be easily reduced in a direct reduction process without the requirement for the blast furnace.

存在使用超高品位铁矿石的可能性，因为即便达不到高炉冶炼要求，这种铁矿石也可以在直接还原工艺中轻松还原。

定语从句 which can be easily reduced in a direct reduction process without the requirement for the blast furnace 被译为原因状语从句，放在主句之后。"这种铁矿石"回指"高品位铁矿石"，在状语从句中用作主语。

请再看几个例子：

Electronic computers which make it possible to free man from the labor of complex measurements and computations have found wide application in engineering.

电子计算机能使人们摆脱花在测量和计算上的复杂劳动，因而在工程技术中已获得广泛应用。（译为原因状语）

Today, pistons in piston-type accumulations ride on Teflon rings which provide longer life and increased reliability.

今天，活塞蓄能器的活塞依靠四氟乙烯环来工作，从而延长了寿命，增加了可靠性。（译为结果状语）

There is a minimum size for the reactor at which the chain reaction will just work.

为了使链式反应刚好能维持下去，反应堆就要有个最起码的尺寸。（译为目的状语）

An electrical current begins to flow through a coil which is connected across a charged condenser.

如果线圈和充电的电容器相连接，电流就开始流过线圈。（译成条件状语）

Iron, which is not so strong as steel, finds wide application.

虽然铁的强度不如钢，但它有广泛的用途。（译成让步状语）

Electricity which is passed through the thin tungsten wire inside the bulb makes the wire very hot.

当电通过灯泡里的细钨丝时，就会使钨丝变得很热。①（译成时间状语）

总之，英语定语从句译成汉语后，其对应成分位置灵活，功能多样。有志于从事科技翻译的人士必须多实践、多揣摩，如此方能万法尽归无法，无法自成万法。

---

① 上述六个例子，均出自：蒲筱梅. 科技英语定语从句汉译的句法重构[J]. 云南农业大学学报，2010 (8)：91-93.

◎ 练习六　请将下面的句子翻译成汉语。

1. In the blast furnace, iron ores such as titaniferous sands, which are rich in titanium, create problems because carbides and nitrides form as solid particles in the slag.

2. Direct reduction processes use gas or coal to effect the reduction of iron ore to solid sponge iron, instead of the hot molten metal which derives from the blast furnace.

3. The iron ore, which may be either totally pelletized or a mixture of pellets and lumps, is fed to the top of the furnace through a conical hopper.

4. In the first step, the iron ore charge is heated and prereduced using reductant gas coming from another reactor in which the reduction itself takes place.

5. The cooling gas is introduced at the bottom of the cooling zone and leaves at the top, where it is passed through a scrubber for cooling and cleaning.

6. The dominant alternative technologies are based on direct reduction and smelting, which emerged during the late 1960s and early 1970s, and in which iron oxide feedstocks are reduced to metallic iron by reducing gases, often at temperatures below the melting point of iron itself, and avoiding the use of coke.

7. As the construction of the vertical shaft furnace is simple, existing and proven designs with a minimum of modification can be used, which also makes it attractive.

8. The new alternative route, of which there are still no commercial examples, tries to convert iron minerals to steel in one reactor in which the four steps of gasification, reduction, smelting and refining all take place.

9. The reductant gas at a temperature of 800 to 900℃ is fed to the reduction furnace and flows counter-current to the descending iron ore, escaping from the top end from where it is recycled back to the process or is exported to be used for electricity generation or the production of chemicals.

10. Not all the advantages of using direct reduction would apply to every plant because there is a wide variety of direct reduction processes from which to choose, and each has its particular advantages.

第七课　碱性氧气转炉炼钢工艺

> 始知百炼钢，永与金刀坚。
>
> ——（明）林鸿《斩蛇剑》

一、科技术语

basic oxygen steelmaking（BOS） A method of primary steelmaking in which carbon-rich molten pig iron is made into steel in the converter. With BOS, oxygen is blown through molten pig iron to lower the carbon content of the alloy and change pig iron into low-carbon steel.（碱性氧气转炉炼钢）

charging aisle A passageway in a steel-making plant close to the converter and used for charging the converter.（装料跨）

top blowing A kind of advanced oxygen blowing furnace steelmaking method. The oxygen top blown converter steelmaking process has the advantages of fast smelting speed, more steel types, better quality and fast construction speed.（氧气顶吹）

bottom blowing A method of producing steel or medium carbon ferromanganese by bottom blowing oxygen converter.（氧气底吹）

refractory lining The refractory layer built in the metal shell of a furnace. The main function is to provide durable containers for high temperature metallurgical melt to complete the ironmaking or steelmaking reaction. It is required that the lining material can withstand high temperature and drastic fluctuation of temperature, and the chemical erosion of slag can resist mechanical impact and wear of molten steel.（耐火炉衬）

continuous casting The process whereby molten metal is solidified into a "semi-finished" billet, bloom, or slab for subsequent rolling in the finishing mills. Prior to the introduction of continuous casting in the 1950s, steel was poured into stationary molds to form ingots.（连铸）

open hearth steelmaking The process in which raw materials such as pig iron and scrap are melted and refined into liquid steel by means of an open hearth furnace, fueled by gas or heavy oil, under the condition of direct heating by a burning flame. (平炉炼钢)

basic oxygen furnace (BOF) A furnace for steelmaking that blows high pure oxygen into the furnace body formed by basic furnace lining by inserting vertical oxygen lance into the furnace mouth to reach the molten pool liquid level. (碱性氧气转炉)

decarburization The process opposite to carburization, namely the reduction of carbon content. The term is typically used in metallurgy, describing the reduction of the content of carbon in metals (usually steel). Decarburization occurs when the metal is heated to temperatures of 700℃ or above when carbon in the metal reacts with gases containing oxygen or hydrogen. (脱碳)

slag splashing The process of using the steelmaking end slag with saturated or supersaturated MgO content through high pressure blowing and splashing to form a high melting point slag layer so as to protect the furnace lining. The formation of slag splashing layer has good corrosion resistance, inhibits oxidation decarburization on the surface of the furnace lining brick, and reduces the high temperature slag brick lining erosion scour. (溅渣护炉)

static charge model A program in an on-site steelmaking process computer that uses initial and final information about the heat (e.g. the amount of hot metal and scrap, the aim carbon and temperature) to calculate the amount of charge and the amount of oxygen required. (静态装料模型)

二、英语原文

The Basic Oxygen Steelmaking (BOS) Process

John Stubbles (Steel Industry Consultant)

Introduction

1 Accounting for 60% of the world's total output of crude steel, the Basic Oxygen Steelmaking (BOS) process is the dominant steelmaking technology. In the U.S., that figure is 54% and slowly declining due primarily to the advent of the "Greenfield" electric arc furnace (EAF) flat-rolled mills. However, elsewhere its use is growing.

2 There exist several variations on the BOS process: top blowing, bottom blowing, and a combination of the two. This study will focus only on the top blowing variation.

3 The Basic Oxygen Steelmaking process differs from the EAF in that it is autogenous, or self-sufficient in energy. The primary raw materials for the BOP are 70%–80% liquid hot metal

from the blast furnace and the balance is steel scrap. These are charged into the Basic Oxygen Furnace (BOF) vessel. Oxygen (>99.5% pure) is "blown" into the BOF at supersonic velocities. It oxidizes the carbon and silicon contained in the hot metal liberating great quantities of heat which melts the scrap. There are lesser energy contributions from the oxidation of iron, manganese, and phosphorus. The post combustion of carbon monoxide as it exits the vessel also transmits heat back to the bath.

4 The product of the BOS is molten steel with a specified chemical analysis at 2900 °F—3000 °F. From here it may undergo further refining in a secondary refining process or be sent directly to the continuous caster where it is solidified into semi-finished shapes: blooms, billets, or slabs.

5 "Basic" refers to the magnesia (MgO) refractory lining which wears through contact with hot, basic slags. These slags are required to remove phosphorus and sulfur from the molten charge.

6 BOF heat sizes in the U.S. are typically around 250 tons, and tap-to-tap times are about 40 minutes, of which 50% is "blowing time". This rate of production made the process compatible with the continuous casting of slabs, which in turn had an enormous beneficial impact on yields from crude steel to shipped product, and on downstream flat-rolled quality.

Basic Operation

7 BOS process replaced open hearth steelmaking. The process predated continuous casting. As a consequence, ladle sizes remained unchanged in the renovated open hearth shops and ingot pouring aisles were built in the new shops. Six-story buildings are needed to house the Basic Oxygen Furnace (BOF) vessels to accommodate the long oxygen lances that are lowered and raised from the BOF vessel and the elevated alloy and flux bins. Since the BOS process increases productivity by almost an order of magnitude, generally only two BOFs were required to replace a dozen open hearth furnaces.

8 Some dimensions of a typical 250 ton BOF vessel in the U.S. are: height 34 feet, outside diameter 26 feet, barrel lining thickness 3 feet, and working volume 8,000 cubic feet. A control pulpit is usually located between the vessels. Unlike the open hearth, the BOF operation is conducted almost "in the dark" using mimics and screens to determine vessel inclination, additions, lance height, oxygen flow, etc.

9 Once the hot metal temperature and chemical analysis of the blast furnace hot metal are known, a computer charge models determine the optimum proportions of scrap and hot metal, flux additions, lance height and oxygen blowing time.

10 A "heat" begins when the BOF vessel is tilted about 45 degrees towards the charging aisle and scrap charge (about 25% to 30% of the heat weight) is dumped from a charging box into the mouth of the cylindrical BOF. The hot metal is immediately poured directly onto the scrap from a transfer ladle. Fumes and kish (graphite flakes from the carbon saturated hot metal) are emitted from the vessel's mouth and collected by the pollution control system. Charging takes a couple of minutes. Then the vessel is rotated back to the vertical position and lime/dolomite fluxes are dropped onto the charge from overhead bins while the lance is lowered to a few feet above the bottom of the vessel. The lance is water-cooled with a multi-hole copper tip. Through this lance, oxygen of greater than 99.5% purity is blown into the mix. If the oxygen is lower in purity, nitrogen levels at tap become unacceptable.

11 As blowing begins, an ear-piercing shriek is heard. This is soon muffled as silicon from the hot metal is oxidized forming silica, SiO_2, which reacts with the basic fluxes to form a gassy molten slag that envelops the lance. The gas is primarily carbon monoxide (CO) from the carbon in the hot metal. The rate of gas evolution is many times the volume of the vessel and it is common to see slag slopping over the lip of the vessel, especially if the slag is too viscous. Blowing continues for a predetermined time based on the metallic charge chemistry and the melt specification. This is typically 15 to 20 minutes, and the lance is generally preprogrammed to move to different heights during the blowing period. The lance is then raised so that the vessel can be turned down towards the charging aisle for sampling and temperature tests. Static charge models however do not ensure consistent turndown at the specified carbon and temperature because the hot metal analysis and metallic charge weights are not known precisely. Furthermore, below 0.2% C, the highly exothermic oxidation of iron takes place to a variable degree along with decarburization. The "drop" in the flame at the mouth of the vessel signals low carbon, but temperature at turndown can be off by +/- 100 °F.

12 In the past, this meant delays for reblowing or adding coolants. Today, with more operating experience, better computer models, more attention to metallic input quality, and the availability of ladle furnaces that adjust for temperature, turndown control is more consistent. In some shops, sublances provide a temperature-carbon check about two minutes before the scheduled end of the blow. This information permits an "in course" correction during the final two minutes and better turn-down performance. However, operation of sublances is costly, and the required information is not always obtained due to malfunctioning of the sensors.

13 Once the heat is ready for tapping and the preheated ladle is positioned in the ladle car under the furnace, the vessel is tilted towards the tapping aisle, and steel emerges from the taphole in the upper "cone" section of the vessel. The taphole is generally plugged with material

that prevents slag entering the ladle as the vessel turns down. Steel burns through the plug immediately. To minimize slag carryover into the ladle at the end of tapping, various "slag stoppers" have been designed. These work in conjunction with melter's eyeballs, which remain the dominant control device. Slag in the ladle results in phosphorus reversion, retarded desulfurization, and possibly "dirty steel". Ladle additives are available to reduce the iron oxide level in the slag but nothing can be done to alter the phosphorus.

14 After tapping steel into the ladle, and turning the vessel upside down and tapping the remaining slag into the "slag pot", the vessel is returned to the upright position. In many shops residual slag is blown with nitrogen to coat the barrel and trunion areas of the vessel. This process is known as "slag splashing". Near the end of a campaign, gunning with refractory materials in high wear areas may also be necessary. Once vessel maintenance is complete the vessel is ready to receive the next charge.

Basic Chemistry and Heat Balance

15 A heat size of 250 tons is used as the basis for the following calculations. This is close to the average heat size for the 50 BOFs which were operable in the U.S. in 1999. The following charge chemistry is assumed (See Table 7.1):

Table 7.1 Charge Chemistry

%C	%Si	%Mn	%S	%P	%Al	Residuals
Hot metal	4.5	.75	1.0	.01	.05	0
Scrap	.05	.05	.4	.015	.01	.03

16 Table 7.1 illustrates the heat balance PER TON OF HOT METAL. It assumes a 75% hot metal in a total charge of 275 tons which yields 250 tons of liquid steel (without alloys). If the oxygen were supplied as air, the heat required to take N_2 from room temperature to 2,900 °F would be about 500,000 Btu per NTHM, which illustrates that the BOS is a Bessemer process with cold scrap substituted for cold nitrogen. (NTHM one short ton or 2,000 pounds of hot metal).

Table 7.2 Heat Balance Per Net Ton of Hot Metal 75% Hot Metal in Charge

HEAT AVAILABLE	Btu (000's)	HEAT REQUIRED	Btu (000's)
C→CO	366	H.M 2,400→2,900 F	220

续表

HEAT AVAILABLE	Btu (000's)	HEAT REQUIRED	Btu (000's)
Si → SiO$_2$	204	FLUXES →2,900 F	110
Mn →MnO	60	O$_2$→2,900 F	120
P →P$_2$O$_5$	10	HEAT LOSSES	50
Fe→FeO	110	SCRAP→2,900 F	415
CO→CO$_2$	130		
SLAG FORMATION	35		
TOTAL	915		915

17 The actual percentage of hot metal in the charge is very sensitive to the silicon content and temperature of the hot metal and obviously increases as these decrease.

18 The oxygen required per heat is shown in Table II, as #/NTHM and as a percentage for the various reactions. 181#/NTHM corresponds to about 18.6 tons/per heat or 1800 scf/tapped ton. Oxygen consumption increases if end-point control is poor and reblows are necessary.

Table 7.3　Oxygen Requirements Per NTHM

REACTION	#/NTHM	% OF TOTAL
C →CO	120	66
Si→SiO$_2$	17	9
Fe→FeO (SLAG)	16	9
CO→CO$_2$	12	7
Fe→FeO (FUME)	8	4
Mn,P→MnO,P$_2$O$_5$	7	4
DISSOLVED OXYGEN	1	1
	181	100

19 The final calculation for yield losses is shown in TABLE III. The metalloids and Mn are oxidized out of the hot metal, the scrap is often coated with Zn which volatilizes, and iron units are lost to the slag, fume, and slopping. To tap 250 tons of liquid steel, 250/0.91 or 275 charge tons are required, of which 206 will be hot metal, and the balance scrap.

Table 7.4 Yield Losses in a BOF Heat

LOSS	% CHARGE
METALLOIDS IN HM (6.3%)	4.7
DEBRIS, COATINGS ON SCRAP (2.5 %)	0.6
IRON TO SLAG	2.2
IRON TO FUME	1
IRON TO SLOPPING	0.5
TOTAL	9

Raw Materials

Hot Metal

20   Hot metal is liquid iron from the blast furnace saturated with up to 4.3% carbon and containing 1% or less silicon, Si. It is transported to the BOF shop either in torpedo cars or ladles. The hot metal chemistry depends on how the blast furnace is operated and what burden (iron-bearing) materials are charged to it. The trend today is to run at high productivity with low slag volumes and fuel rates, leading to lower silicon and higher sulfur levels in the hot metal. If BOF slag is recycled, P and Mn levels rise sharply since they report almost 100% to the hot metal. U.S. iron ores are low in both elements.

21   The sulfur level from the blast furnace can be 0.05% but an efficient hot metal desulfurizing facility ahead of the BOF will reduce this to below .01%. The most common desulfurizing reagents, lime, calcium carbide and magnesium—used alone or in combination—are injected into the hot metal through a lance. The sulfur containing compounds report to the slag; however, unless the sulfur-rich slag is skimmed before the hot metal is poured into the BOF, the sulfur actually charged will be well above the level expected from the metal analysis.

Scrap

22   In autogenous BOS operation, scrap is by far the largest heat sink. At 20%-25% of the charge it is one of the most important and costly components of the charge.

23   Steel scrap is available in many forms. The major categories are "home scrap", generated within the plant. With the advent of continuous casting, the quantity of home scrap has diminished and it is now necessary for integrated mills to buy scrap on the market. Flat rolled

scrap is generally of good quality and its impact on the chemistry of BOF operations can almost be ignored. There is a yield loss of about 2% due to the zinc coating on galvanized scrap. "Prompt scrap" is generated during the manufacturing of steel products. It finds its way into the recycling stream very quickly. Many steel mills have agreements with manufacturers to buy their prompt scrap. "Obsolete" or "post-consumer" scrap returns to the market after a product has ended its useful life. Cans return to the market very quickly but autos have an average life of 12 years.

24 Scrap also comes in many sizes, varying chemical analyses and a variety of prices, all of which makes the purchase and melting of scrap a very complex issue. Very large pieces of scrap can be difficult to melt and may damage the vessel when charged. Some scrap may contain oil or surface oxidation. Obsolete scrap may contain a variety of other objects which could be hazardous or explosive. Obviously the chemical analysis of obsolete scrap is imprecise.

25 Scrap selection is further complicated by the wide variety of steel products. Deep drawing steels limit the maximum residual (%Cu +%Sn + %Ni +%Cr +%Mo) content to less than 0.13%, while other products allow this to range as high as 0.80%. Since these elements cannot be oxidized from the steel, their content in the final product can only be reduced by dilution with very high purity scrap or hot metal. The use of low residual hot metal in the BOS, with its inherent dilution effect, is one of the features that distinguish BOF from EAF steelmaking.

Fluxes

26 Fluxes serve two important purposes. First they combine with SiO_2 which is oxidized from the hot metal to form a "basic" slag that is fluid at steelmaking temperatures. This slag absorbs and retains sulfur and phosphorus from the hot metal.

27 Lime (95+% CaO) and dolomite (58%CaO, 39% MgO) are the two primary fluxes. They are obtained by calcining the carbonate minerals, generally offsite in rotary kilns. Calcining $CaCO_3$ and $MgCO_3$ liberates CO_2 leaving CaO or MgO. Two types, "soft" and "hard" burned lime, are available. A lump of soft burned lime dissolves quickly in a cup of water liberating heat. Hard burned material just sits there. Soft burned fluxes form slag more quickly than hard-burned, and in the short blowing cycle, this is critical for effective sulfur and phosphorus removal. The amount of lime charged depends on the Si content of the hot metal.

28 In BOS steelmaking a high CaO/SiO_2 ratio in the slag is desirable, e.g. A rule of thumb is 6 X the weight of Si charged. The MgO addition is designed to be about 8% to 10% of the final slag weight. This saturates the slag with MgO, thus reducing chemical erosion of the

MgO vessel lining.

Coolants

29 Limestone, scrap, and sponge iron are all potential coolants that can be added to a heat that has been overblown and is excessively hot. The economics and handling facilities dictate the selection at each shop.

Alloys

30 Bulk alloys are charged from overhead bins into the ladle. The common alloys are ferromanganese (80% Mn, 6% C, balance Fe), silicomanganese (66% Mn, 16% Si, 2% C, balance Fe), and ferrosilicon (75% Si, balance Fe). Aluminum can be added as shapes and/or injected as rod. Sulfur, carbon, calcium, and special elements like boron and titanium are fed at the ladle furnace as powders sheathed in a mild steel casing about 1/2 inch in diameter.

Refractories

31 The basis for most refractory bricks for oxygen steelmaking vessels in the U.S. today is magnesia, MgO, which can be obtained from minerals or seawater. Only one dolomite (MgO + CaO) deposit is worked in the U.S (near Reading, PA). For magnesia, the lower the boron oxide content, and the lower the impurity levels, the greater the hot strength of the brick. Carbon is added as pitch (tar) or graphite.

32 The magnesia lime type refractories used in lining oxygen steelmaking vessels are selected mainly for their compatibility with the highly basic finishing slags required to remove and retain phosphorus in solution. During refining, the refractories are exposed to a variety of slag conditions ranging from 1 to 4 basicity as silicon is oxidizes from the bath and combines with lime. The iron oxide, FeO, content of the bath increases with blowing time especially as the carbon in the steel falls below 0.2 % and Fe is oxidized. Although all refractory materials are dissolved by FeO, MgO forms a solid solution with FeO, meaning they coexist as solids within a certain temperature range. The high concentrations of FeO formed late in the blow, however, will oxidize the carbon in the brick.

33 The original bricks were tar bonded. The MgO grains were coated with tar and pressed warm, which represented a great step forward for the BOS process. Tempering removed volatiles. In service, the tar was coked and the residual intergranular carbon resisted slag wetting and attack by FeO. In addition, as the tar softened during vessel heat-up, the lining was relieved of expansive stresses. Hot strength was increased by sintering bricks made from pure MgO grains

at a high temperature and then impregnating them with tar under a vacuum. However, for environmental reasons these types of bricks are no longer used in oxygen steelmaking.

34 Today's working lining refractories are primarily resin-bonded magnesia-carbon bricks made with high quality sintered magnesite and high purity flake graphite. Resin-bonded brick are unfired and contain 5% to 25% high purity flake graphite and one or more powdered metals. These bricks require a simple curing step at 350 to 400°F to "thermoset" the resin that makes them very strong and therefore easily handled during installation. Further refinements include using prefused grains in the mix. Small additions of metal additives (Si, Al, and Mg) protect the graphite from oxidation because they are preferentially oxidized. Metallic carbides, nitrides, and magnesium-aluminate spinel form in service at the hot face of the brick filling voids, and adding strength and resistance to slag attack.

35 The rate of solution of a refractory by the slag is dependent on its properties. These properties are directly related to the purity and crystal sizes of the starting ingredients as well as the manufacturing process. Additions of up to 15% high purity graphite to MgO-carbon refractories provide increased corrosion resistance. Beyond 15% this trend is reversed due to the lower density of the brick. Ultimately, the cost per ton of steel for brick and gunning repair materials, coupled with the need for vessel availability, dictate the choice of lining.

36 The penetration of slag and metal between the refractory grains, mechanical erosion by liquid movement, and chemical attack by slags all contribute to loss of lining material. Over the years, there have been numerous operating developments designed to counteract this lining wear:

37 i) Critical wear zones (impact and tap pads, turndown slag lines, and trunion areas) in furnaces have been zoned with bricks of the highest quality.

38 ii) "Slag splashing" whereby residual liquid slag remaining after the tap is splashed onto the lining with high pressure nitrogen blown through the oxygen lance. This seemingly simple practice has increased lining life beyond all expectations, from a few thousand to over 20,000 heats per campaign.

39 iii) Instruments are now available to measure lining contours in a short time period, to maximize gunning effectiveness using MgO slurries.

40 iv) Dolomite (40%MgO) is added to the flux addition to create slags with about 8% MgO, which is close to the MgO saturation level of the slag.

41 v) Improved end-point control resulting in lower FeO levels and shorter oxygen-off to charge intervals have reduced refractory deterioration.

42 None of the above would be significant however, without the improvements in quality

and type of basic brick available to the industry.

43 Today, the refractory industry is undergoing major structural changes. Companies are being continually acquired and the total number of North American suppliers is greatly reduced. A very high percentage of refractory materials are being produced off shore, with China being the most significant newcomer.

Environmental Issues

44 Environmental challenges at BOS shops include: (1) the capture and removal of contaminants in the hot and dirty primary off-gas from the converter; (2) secondary emissions associated with charging and tapping the furnaces; (3) control of emissions from ancillary operations such as hot metal transfer, desulfurization, or ladle metallurgy operations; (4) the recycling and/or disposal of collected oxide dusts or sludges; and (5) the disposition of slag.

45 In the U.S., most BOF primary gas handling systems are designed to generate plant steam from the water-cooled hood serving the primary system. About half of the systems are open combustion designs where excess air is induced at the mouth of the hood to completely burn the carbon monoxide. The gases are then cooled and cleaned either in a wet scrubber or a dry electrostatic precipitator. The remainder of U.S. systems are suppressed combustion systems where gases are handled in an uncombusted state and cleaned in a wet scrubber before being ignited prior to discharge. In both cases, the cleaned gases must meet EPA-mandated levels for particulate matter.

46 Suppressed combustion systems offer the potential for recovery of energy, a practice that is more prevalent in Europe and Japan. However, in the U.S., other than steam generation, no attempt is made to capture the chemical or sensible heat in the off-gas leaving the vessel. While this represents the loss of a considerable amount of energy (about 0.7 million Btu/ton), the pay-back on capital required, either for the conversion of open combustion to suppressed combustion systems or the addition of necessary gas collection facilities for suppressed combustion systems, is over 10 years. In addition, the necessity of taking shops out of service to make these changes is not practical. Most BOF shops in the U.S. pre-date the energy crises of the 1970s, and even today, energy in the U.S. is relatively less expensive than it is abroad.

47 Secondary fugitive emissions associated with charging and tapping the BOF vessel, or emissions escaping the main hood during oxygen blowing, may be captured by exhaust systems serving local hoods or high canopy hoods located in the trusses of the shop or both. Typically, a fabric collector, or baghouse, is used for the collection of these fugitive emissions. Similarly, ancillary operations such as hot metal transfer stations, desulfurization, or ladle metallurgy

operations are usually served by local hood systems exhausted to fabric filters.

48　The particulate matter captured in the primary system, whether in the form of sludge from wet scrubbers or dry dust from precipitators, must be processed before recycling. Sludge from wet scrubbers requires an extra drying step. Unlike EAF dust, BOF dust or sludge is not a listed hazardous waste. If the zinc content is low enough, it can be recycled to the blast furnace or BOF vessel after briquetting or pelletizing. Numerous processes for recycling the particulate are in use or under development.

49　BOF slag typically contains about 5% MnO and 1% P_2O_5 and are often can be recycled through the blast furnace. Because lime in steel slag absorbs moisture and expands on weathering, its use as an aggregate material is limited, but other commercial uses are being developed to minimize the amount that must be disposed.

Conclusion

50　The BOS has been a pivotal process in the transformation of the U.S. steel industry since World War Ⅱ. Although it was not recognized at the time, the process made it possible to couple melting with continuous casting. The result has been that melt shop process and finishing mill quality and yields improved several percent, such that the quantity of raw steel required per ton of product decreased significantly.

51　The future of the BOS depends on the availability of hot metal, which in turn depends on the cost and availability of coke. Although it is possible to operate BOFs with reduced hot metal charges, i.e. < 70%, there are productivity penalties and costs associated with the supply of auxiliary fuels. Processes to replace the blast furnace are being constantly being unveiled, and the concept of a hybrid BOF-EAF is already a reality at the Saldahna Works in South Africa. However, it appears that the blast furnace and the BOS will be with us for many decades into the future.

三、分析与讲解

高炉生产的铁水含碳量高，制成的生铁硬而脆，韧性差，焊接性能也差，几乎没有塑性，不能进行轧制、锻压等塑性变形加工。此外，生铁中含硫、磷等杂质多，应用范围受限，必须用氧化方法去除生铁中的杂质，同时通过加入适量的合金元素，使之成为具有高强度、高韧性或其他特殊性能的钢。①

---

①　铁和钢最根本的区别在于含碳量。通常，含碳量小于 0.02% 的称为熟铁，在 0.02%—2.11% 之间的称为钢，在 2.11%—4.3% 之间的称为生铁。

炼钢技术从 19 世纪开始，经历了贝塞麦炼钢炉（Bessemer converter）、平炉（open-hearth furnace）、氧气炼钢炉（oxygen steelmaking furnace）和电弧炉（electric arc furnace）几个主要阶段。1856 年，英国人贝塞麦（Henry Bessemer）发明了底吹空气的酸性转炉炼钢工艺。空气进入炉底气室，然后通过透气砖进入钢液熔池，但该技术因为酸性内衬而无法冶炼高磷铁水。1879 年，英国人托马斯（Sidney Gilchrist Thomas）发明了碱性耐火材料和碱性炉渣的底吹空气转炉炼钢技术。19 世纪 60 年代，生活在英国的德国人西门子（Karl Wilhelm Siemens）和法国人马丁（Pierre-Emile Martin）分别发明了在高温蓄热室结构的炉子内使用铁矿石为氧化剂实现铁液脱碳的炼钢过程，即"西门子—马丁炉炼钢工艺"（Siemens-Martin Process），又称为平炉炼钢技术。1949 年，瑞士人杜勒（Robert Durrer）成功完成了顶吹氧气转炉炼钢的试验，称为 LD 工艺①。随后，该工艺在欧美等发达国家广泛推广，在美国被称为氧气碱性转炉炼钢技术，简称 BOF 工艺。20 世纪 60 年代，底吹氧气炼钢技术和顶底复吹技术开始出现并逐渐推广。19 世纪 70 年代，西门子建造了第一座试验用的炼钢电弧炉。20 世纪初，美国建造了世界上第一座电弧炉并开启了电弧炉炼钢实践。②

本篇文章是对碱性氧气转炉炼钢的简单介绍。以下是对文章的分析和讲解：

第 1 段：(1)第 1 句中，dominant 在原文中作定语，翻译时不妨转换为动词谓语"处于主导地位"。如此处理，主要是因为汉语中"是"字判断句用得不如英语频繁，形容词译成动词谓语，更符合汉语的表达规律。从语序上来说，翻译第 1 句时，应根据事理逻辑顺序，采用倒译法，先翻译主谓结构，再译现在分词短语 Accounting for 60% of the world's total output of crude steel。(2)第 2 句中，that figure 应译成"这个比例"，而不是"那个数字"。后面的 Greenfield，因为首字母是大写，所以一定要确定这个词是专有名词还是普通名词。经查阅各种参考文献，可以确定虽然 G 大写，但这里的 Greenfield 事实上是个普通名词，意思为"新建的"。

英语、汉语在近指、远指的表达上有相同之处，也有不同之处，具体如下：

指称时间	过去（较远）	过去（较近）	现在	将来（较近）	将来（较远）
英语	that	that	this	this	this
汉语	那	这	这	这	那

例如：

The first known European breakthrough in the production of cast iron, which led

---

① 之所以称为 LD 工艺，因为世界上使用这种工艺最早的两座钢厂分别在奥地利的林茨市（Linz）和多纳维茨市（Donawitz）。

② 张立峰. 炼钢技术的发展历程和未来展望(I)——炼钢技术的发展历程[J]. 钢铁，2022(12)：1.

quickly to the first practical steel, did not come until 1740. In <u>that</u> year, Benjamin Huntsman took out a patent for the melting of material for the production of steel springs to be used in clockmaking.

已知欧洲在铸铁生产上的第一次突破，直到 1740 年才出现，并很快带来了具有实用价值的钢的首次生产。<u>那一年</u>，本杰明·亨茨曼申请获批了一项熔化材料生产钢弹簧的专利，用于钟表制作。

However, these appearances are deceiving. New ore-enrichment techniques have made the use of lower-grade ore much more attractive, and there is a vast supply of <u>that</u> ore.

然而，<u>这些</u>表象带有欺骗性。新研发的矿石富集技术让低品位矿石的使用变得更有吸引力，而<u>这种</u>矿石的供应量很大。

这里的两个例句，前一个例句中 that year 指代较远的 1740 年，所以译成"那一年"，后一个例句中的 that ore 指代前一句刚刚提到的 lower-grade ore，所以不能译成"那种矿石"，必须译成"这种矿石"——当然重复"低品位矿石"也是可以的。

第 2 段：(1) BOS 缩写可以不翻译，因为汉语冶金科技文本中有时也直接使用英文缩写。其他的一些常用的缩写也可以如此处理，如 EAF、BOF。当然翻译成汉语也没问题。(2) top blowing, bottom blowing 都是指 blowing of oxygen，直译成"顶吹""底吹"也可以，加上"氧气"两个字（"氧气顶吹""氧气底吹"）意思更清楚。a combination of the two 是"顶底复合吹氧"。第 2 句中的 the top blowing variation，如果译成"顶吹类"，就显得不专业，不如译成"顶吹炼钢"。

第 3 段：(1) 第 2 句中的 the balance 意思是 something that is left over，这里应译为"差额部分"或"不足部分"。(2) 第 5 句是个存在句，按原语序不好翻译，可以将 the oxidation of iron, manganese, and phosphorus 译成主语。将介词宾语译成主语，是存在句常用的英汉翻译技巧。还比如：

There exist disputes in transport pathway and transport mechanism of electrical signals.
电信号的传递途径及传递机制存在争议。
There exist differences in the dry matter ratio of each organ under different circumstances.
各器官干物质率在不同状态下存在着不同程度的差异。
There are still a few residual problems with the computer program.
电脑程序还有一些残留问题。

第 4 段：为方便中国读者理解，应加注 2900 ℉—3000 ℉对应的摄氏温度"1593.3 ℃—

1648.9℃"放在括号中。

第6段：翻译第1句时要注意译句语言的精练、简洁。如按原句翻译，译文便是"在美国，BOF 炉的容积通常约为 250 吨，炼钢到出钢时间大约为 40 分钟，其中一半为吹炼时间"。不如译成："在美国，典型 BOF 炉每炉产钢约 250 吨，需时约 40 分钟，其中一半为吹炼时间"。

第7段：(1) 第2句 predated 是"(在日期上) 早于；先于"的意思，但如果译成"时间上早于连铸"，容易让读者误解以为前面 replaced open hearth stelemaking 这个动作在时间上早于连铸，所以要说清楚："在工序上处在连铸之前"。(2) 第4句翻译时，须先译目的状语，然后才译主句。accommodate 可以译为"装下"或"容纳"；long oxygen lances 最好译成"长长的氧枪"，而不是"长的氧枪"或"长氧枪"，读来声音上更和谐；elevated 有两种译法："升高的"或"抬高的/抬升的"，前者暗示上面有东西拉升，后者暗示下面有可活动的底座，这里应译成"抬高的/抬升的"。如果不确定是拉升还是抬升，可模糊译为"高位"。(3) 最后一句中的 a dozen 不要直译成"十二座"，而要模糊译成"十几座"，因为 BOF 炉和平炉的炉容不是固定不变的，所以不可能刚好是"十二座"。

汉语中，表达同样的意思，有时有单字词语和双字词语可以选择，一般说来，单字词语较口语化，双字词语较正式。例如："要"和"为了"。To accommodate the long oxygen lances that are lowered and raised from the BOF vessel and the elevated alloy and flux bins 可以译为"要装下可从 BOF 转炉升降的长长的氧枪及抬高的合金和熔剂料仓……"，也可以译为"为了装下可从 BOF 转炉升降的长长的氧枪及抬高的合金和熔剂料仓……"，第二种译法更正式一些。

类似的词语对还有："读"和"阅读"，"书"和"书籍"，"扫"和"清扫"，"洗"和"洗涤"，"买"和"购买"，"加"和"添加"，等等。

第9段：翻译 flux additions, lance height and oxygen blowing time 都应添加词语："熔剂(应)添加量、氧枪的(适宜)高度和(最佳)氧吹时间"，将原表达中暗含的意思显性化。

第10段：(1) 第1句中两个 heat，第一个 heat 应译成"炉"，第二个 heat 要译成"炉料"。科技翻译中，也常常需要将同一个词语翻译成不同的意思。(2) The lance is water-cooled with a multi-hole copper tip 描述氧枪，直译的话很别扭("氧枪是水冷，带有一个多孔的铜制前端")，需要放弃原句判断句句式，采用汉语流水句："氧枪采用水冷方式，前端为铜制，上有多个喷孔"。

第11段：(1) 第1、2句都在描述吹氧时发声的情况，可合译成一句话，描述声音前后的变化。第2句中的 as 引导的从句是 This is soon muffled 的原因或者说先决条件，所以应该先译，然后才译 This is soon muffled。(2) the melt specification 的字面意思是"熔体的规格"，其实就是期望的"钢的规格"。(3) the highly exothermic oxidation of iron takes place to a variable degree 字面意思是"高度放热的铁的氧化发生的程度有所不同"，真实意思是"铁的氧化程度会有不同，释放的热量也有变化"，结合所在语境，可译为"铁的氧化释放的大量热量会有所变化"。

第 12 段：(1) temperature-carbon check 直译为"温度和碳含量检测"也可以，但不如"测温定碳"言简意赅。"in course correction"意为"边冶炼边调节"，不妨采用正说反译，将其译为"不停炉调节"。

第 13 段：These work in conjunction with melter's eyeballs, which remain the dominant control device 句中的 melter's eyeballs 是借代，用 eyeballs 指代炼钢工人的观察，翻译时可隐性化处理："这些挡渣器在炼钢工人手里成为了主要的控制设备"。

第 14 段：第 4 句中 gunning with refractory materials in high wear areas may also be necessary 主语太长，顺着翻译会导致头重脚轻，可以逆着翻译为"可能有必要对转炉炉衬磨损严重的地方喷补耐火材料"。

第 15 段：第 1、2 句关系紧密，第 2 句中的 This 就指第 1 句中的 A heat size of 250 tons，所以两句可合译成一句。A heat size of 250 tons 与 the average size for the 50 BOFs which were operable in the U. S. in 1999 之间的关系更为紧密，而与 is used as the basis for the following calculations 之间的关系相对较弱，考虑到这一点，不妨将第 2 句的译文放在第 1 句译文中，两边用破折号隔开："以 250 吨炉容——这跟美国 1999 年在运转的 50 座 BOF 的平均炉容接近——为基础进行下述计算"。

第 16 段：汉语中总是先给出假设条件，然后再叙述在假设条件下产生的结果，所以翻译时应先译第 2 句，然后再译第 1 句。

第 17 段：既可以直译成"炉料中铁水的实际比例对于铁水的硅含量和温度十分敏感，明显随着后两者的降低而增加"，也可以意译为"炉料中铁水的实际比例受铁水的硅含量和温度影响很大，明显随着后两者的降低而增加"。

第 18 段：翻译 as #NTHM and as a percentage for the various reaction 时，须结合 Table 7-3 来斟酌。可以看到，在 Table 7-3 中，#/NTHM 和% OF TOTAL 分别是后两栏的标题，因此 as #NTHM and as a percentage for the various reaction 可译成"按'#/每吨铁水'和'不同反应需氧占比'两栏列出"。

科技英语中图片和表格较多。翻译与图片、表格相关的文字时，一定要结合图片、表格内容再作决定。

第 20 段：第 5 句中的 report 这种用法比较少见，在这里是 present oneself 的意思，可译成"回到"。因为 BOF 炉渣中磷和锰较多，所以如果 BOF slag is recycled，那么炉渣中的磷和锰也将几乎 100% 回到铁水中。

第 21 段：英语是形合语言，为树枝型结构。原文中，第二句的主干为 The most common desulfurizing reagents... are injected into the hot metal through a lance，而 lime, calcium carbide and magnesium 为"desulfurizing reagents"的同位语，used alone or in combination 为插入语，是对 desulfurizing reagents 的补充说明。汉语是意合语言，为竹节型结构，所以不妨采取顺译法，按照逻辑顺序断句为 The most common desulfurizing reagents, lime, calcium carbide and magnesium// —used alone or in combination —//are injected into the hot metal through a lance，然后依次译出，符合汉语流水句的特点。

第 22 段：the largest heat sink 字面意思是"最能让热量下降"，也就是"最吸热"。

第 23 段：(1) home scrap 有"本厂废钢"和"自产废钢"两种译法，任选一译法即可。(2) 最后一句中的 Cans，当然指"钢制的罐头"，而不是"铝制的罐头"，所以不能简单地译为"罐头"。

有些科技术语的翻译没有得到规范，而是有几种译法，比如这里的 home scrap 及后面的 Prompt scrap（"边角料废钢"或"加工废钢"）、"Obsolete" scrap（"折旧废钢"或"报废产品废钢"或"循环废钢"）。在这种情况下，只要选择其中一种译法即可。但需要注意的是，一旦选定了一种译法，整个译文都必须采用这种译法，不能一会儿这种译法，一会儿另外一种译法，这样会让读者误解是两个不同的术语。比如这里采用了"自产废钢"的译法，全篇译文都必须使用这种译法，不能变换。

第 25 段：(1) 第 1 句中，the wide variety of steel products 事实上是 Scrap selection is further complicated 的原因，所以应将 by 的宾语译成原因状语，放在句前，然后再译主句："因为钢制产品种类太多，所以废钢的选择就更加复杂。"

第 26 段：第 1、2 句本可以译成精炼的"熔剂的主要作用有二：一是……"，但原文只提到了一种作用，用 First 引出，并没有给出另外一种作用，所以翻译时需要稍加变通，译成"熔剂的主要作用有两个，其中之一是……"。

科技翻译时，不仅需要关注译句本身是否准确、通顺，还要注意放在整个段落和篇章的语境下，译句的表达是否前后连贯。

第 27 段：(1) 第 2 句中，They 可以译为"它们"，但科技汉语中常用"两者"，最好译成"两者"，所以代词的翻译也值得琢磨。后面的 offsite 是指 off the steel plant，也就是"钢厂外"。(2) Hard burned material just sits there 是拟人的手法，指"硬烧石灰不会溶解于水中释放热量"，也就是"硬烧石灰则什么也不会发生"。

第 29 段：翻译第 2 句时，为保持同前句的衔接和连贯，需要增加词语："<u>至于具体到某一炼钢车间应选择何种冷却剂，则取决于成本和设备上</u>的考虑。"

第 30 段：两处的 alloys 都应该译成"合金料"，而不是"合金"，表明 alloys 是作为原材料而被添加进炼钢炉的。

第 32 段：(1) 第 1 句是个长复合句，需要基于意群断句。过去分词短语 required to remove and retain phosphorus in solution 作为后置定语修饰 the highly basic finishing slags, for their compatibility 陈述原因，主语 The magnesia lime type refractories used in lining oxygen steelmaking vessels are selected mainly 是被动句。翻译时，应先译 the highly basic finishing slags required to remove and retain phosphorus in solution，将为什么需要"高碱性炉渣制品"解释清楚，再将主句由被动句译为汉语无主句，指出"选择镁质石灰质耐火材料用作氧气炼钢炉的炉衬"，随后解释如此选择的原因"主要是因为这些耐火材料与需要的高碱性炉渣制品相容"。调整语序后，译文逻辑清晰，表达自然。当然，还有一种译法："选择镁质石灰质耐火材料用作氧气炼钢炉的炉衬，主要是因为这些耐火材料不易受除去、吸纳钢水中磷需要的高度碱性的终渣的侵蚀"，更紧凑，但读来有点不够轻松。(2)第2句结构也比较复杂，应按照如下顺序翻译：During refining // as silicon is oxidized from the bath and combines with lime // ranging from 1 to 4 basicity // the refractories are exposed to a variety of slag conditions。

第33段：倒数第2句各成分之间的关系比较复杂，需要仔细分析：先是在高温下(at a high temperature)烧结(sintering) MgO 颗粒(MgO grains)，然后将颗粒烧结形成的砖(bricks)在真空中浸满焦油(impregnating them with tar under a vacuum)，这样就会增大耐火砖的高温强度(Hot strength was increased)。翻译时也应按照这样的顺序来表述。

第35段：倒数第2句 Beyond 15% this trend is reversed 最好不要直译为"超过15%，这种趋势就会因为砖的较低密度而被逆转"，而应在充分理解的基础上大胆意译："但超过15%的添加量将会适得其反，因为耐火砖的密度更低了"。

第41段：不要基于原句结构进行翻译，而应基于意群，将整个句子译成四个小句，形成流水句："改善终点控制，降低 FeO 的水平，缩短停吹到装料的间隔，减少耐火材料的磨损"，这样处理，不仅表达简练，而且层次清楚。

第42段：英、汉语在用词、造句和表达等方面有诸多不同，翻译时常需要采用增词法。具体到这句，需要增加"如果"，表达出 without the improvements in quality and type of basic brick 中隐含的条件关系，便于目标语读者理解。

第43段：这一段落的三个句子都围绕 refractory industry 展开阐述，可将它们合译成一个由多个并列分句构成的长句，更符合汉语的表达习惯。newcomer 意思是"新来的人；新参加的人"，这里不妨站在作者的视角进一步引申，译为"新竞争对手"。

第44段：原文五个 Environmental challenges，都是用名词词组来表达的，译成汉语的名词短语读来会很不自然，不妨按照汉语的表达习惯，译成"如何+动宾结构"的小句，不仅方便处理，而且形成排比，层次清楚。

第45段：EPA-mandated levels for particulate matter 字面意思是"环境保护局强制执行的颗粒物质水平"，也就是"环境保护局制订的颗粒物质排放强制标准"。

第48段：a listed hazardous waste 中的 listed，指被环保部门登记在册的，也就是"明文规定的"。

第49段：第2句中的 weathering 不好翻译。这里 weather 的意思是 To expose to the

action of the elements in the air，也就是"暴露于空气之中"。

第 50 段：英、汉语语言结构不完全相同，当原文表层结构与汉语表达方式不一致时，需要进行句子结构的调整。第 2 句表示原因，第 3 句表示结果，可以合译成一个"前因后果"的汉语句子。the process 译为"该工艺"比"这个工艺"好，"该"体现了科技文本用语正式的特点。

第 51 段：第 2 句中 productivity penalties 是一种抽象的说法，翻译时需要将其具体化译为"生产率降低"。

英语中抽象名词用得较多，科技英语也是如此，但汉语倾向于使用表示具体现象、动作等的名词，英译汉时，常常需要做这种"抽象"到"具体"的转换工作。比如：

The operation of an electric machine needs some knowledge of its performance.
操作电机需要了解一些电机的性能。

The study of the nature and course of the complex cutting is usually achieved by means of observation, research, and analysis of phenomena and processes using simplified models of the actual cutting process.
研究复杂切削工艺本质及其过程，通常是采用实际切削过程的简化模型，通过对切削现象和过程进行观察、研究和分析来进行。

Tolerance is the amount of variation in a dimension which may be permitted or tolerated without impairing the functional fitness of the part. It is important not to give a tolerance closer than is absolutely necessary.
公差是在不影响零件的配合功能的前提下允许的尺寸变动量。给出的公差不要小于绝对必要的值，这很重要。

四、参考译文

碱性氧气转炉炼钢工艺
钢铁行业顾问 约翰·斯塔布斯

简介

1 碱性氧气转炉炼钢（BOS）工艺在炼钢技术中处于主导地位，生产的粗钢占全世界粗钢总产量的 60%。在美国，这个比例为 54%，且正缓慢下降，主要是由新建电弧炉扁钢轧机厂的投产引起的。然而在其他地区，BOS 工艺的使用在增长。

2　BOS 工艺分以下几种：(氧气)顶吹、(氧气)底吹和顶底复合吹(氧)。本研究将只讨论顶吹炼钢。

3　BOS 工艺同电弧炉炼钢工艺(EAF)不同之处在于，BOS 工艺为自热型，能量自足。碱性氧气转炉炼钢(BOP)的主要原材料是高炉铁水，占 70%—80%，差额部分为废钢。铁水和废钢被装进碱性氧气炼钢炉(BOF)内，同时氧气(纯度超 99.5%)以超音速的速度被吹进炉中。氧气氧化铁水中的碳和硅，释放大量热量，熔化废钢。铁、锰和磷的氧化也会贡献一些热量，但较少。一氧化碳从炉中逃逸时的二次燃烧也会将热量传回熔池。

4　2900 $℉$ ~3000 $℉$ (1593.3℃—1648.9℃)时，BOS 的产物为带有特定化学性质的钢水。钢水离开转炉后可能会经过二次精炼或被直接送到连铸机固化为半成品形状：大方坯、小方坯或板坯。

5　"碱性"指的是氧化镁(MgO)耐火炉衬，它因与高温碱性炉渣接触而受到磨损，但从铁水中去除磷和硫需要这些炉渣。

6　在美国，典型 BOF 每炉产钢约 250 吨，需时约 40 分钟，其中一半为吹炼时间。这样的生产速度很好地配合了板坯的连铸，而后者又对从粗钢到成品(发运的产品)收得率及下游的平轧质量有着巨大的、积极的影响。

基本操作

7　BOS 工艺取代了平炉炼钢，在工序上处在连铸之前。因此，在改造后的平炉车间，钢包大小并没有变化，而在新车间建造了铸锭工段。为了装下从 BOF 转炉升降的长长的氧枪及抬高的合金和熔剂料仓，安放 BOF 转炉的建筑物需要六层楼高。由于 BOS 工艺让生产率增加了近一个数量级，所以通常两座 BOF 就可以代替十几座平炉。

8　一座美国的 250 吨 BOF 转炉的尺寸通常是：高 34 英尺(约 10.4 米)，外直径 26 英尺(约 7.9 米)，炉身内衬厚 3 英尺(约 0.9 米)，工作容积 8000 立方英尺(约 226.5 立方米)。控制台通常位于转炉之间。跟平炉不同的是，BOF 操作基本是"在黑暗中"进行的，使用模拟设备和显示屏来控制炉体的倾斜、炉料的添加、氧枪的高度和氧气的流量等。

9　一旦知道了铁水的温度和高炉铁水的化学成分，计算机装料模型就会计算出废钢和铁水的最佳配比、熔剂(应)添加量、氧枪的(适宜)高度和(最佳)氧吹时间。

10　每一次熔炼开始时，BOF 炉体向装料跨倾斜 45 度，废钢(大约占炉料总重的 25%—30%)从装料箱被装入圆柱形 BOF 转炉的炉口。随后，铁水立即被从铁水罐车直接倒在废钢之上。烟气和片状石墨(即来自饱含碳的铁水的石墨片)从炉口逸出，被污染控制系统收集起来。装料会花去几分钟时间，然后炉体转回垂直状态，石灰石/白云石熔剂从高架料仓投送到炉料上，同时氧枪下降到离炉底只有几英尺的高度。氧枪采用水冷方式，前端为铜制，上有多个喷孔。通过氧枪，纯度在 99.5% 以上的氧气喷入混合炉料中。如果氧气纯度不够，则出钢时钢水中的氮含量不达标。

11 吹氧开始时，会听到刺耳的、尖锐的声音，但很快随着铁水中的硅被氧化形成二氧化硅(SiO_2)，二氧化硅又与碱性熔剂反应形成气熔渣包裹氧枪，刺耳的声音消失。气体主要是从铁水中的碳形成的一氧化碳(CO)。气体析出速率比炉体的容积大很多倍，因此经常可以见到炉渣溢出炉嘴，特别是炉渣太黏时。吹氧的时间根据金属炉料的化学性质和钢的规格事先确定，常为15—20分钟。吹氧期间，氧枪通常根据事先的设计移动到各个不同的高度。然后，氧枪抬起，以便炉体向装料跨倾斜，进行取样和温度测试。然而，静态装料模型却并不能确保达到特定碳含量和温度时持续停吹，因为铁水的化学成分和金属炉料的重量无法准确知道。此外，含碳量低于0.2%时，伴随着脱碳过程发生的铁的氧化释放的大量热量会有所变化。炉嘴火焰"下降"意味着含碳量低，但停吹时的温度离期望值却能相差+/-100℉(37.8℃)。

12 在过去，这意味着必须重新吹氧或添加冷却剂。但今天，随着操作经验的增加、计算机模型的优化、金属炉料质量的提高以及可进行温度调节的钢包炉的使用，氧吹控制更连贯。在有些(炼钢)车间，在计划/预定吹氧结束前两分钟左右使用副枪测温定碳，这样便可以在最后两分钟进行不停炉调节，确保获得更好的停吹效果。然而，使用副枪成本很高，且(可能会)因传感器/探测器失灵得不到所需的信息。

13 一旦一炉钢炼好准备出钢，预热钢包安放在炉下钢包车中，炉体(就)向出钢跨倾斜，钢水从炉体上部"炉帽"处的出钢口流出。出钢口通常用材料塞住，防止炉体向下倾斜时，炉渣进入钢包。钢水立刻烧穿塞子。为了最大程度减少出钢结束时的炉渣进入钢包，人们设计了各种各样的"挡渣器"。这些挡渣器在炼钢工人手里成了主要的控制设备。钢包中的炉渣会让磷重新回到钢水中，延迟脱硫过程，可能产生"不洁钢"。已有钢包添加剂可以降低渣中的铁氧化物水平，但对于磷则没有什么办法。

14 出钢进钢包后，炉体被翻转向下，剩余炉渣被倒进"渣罐"，(然后)炉体转回直立状态。在很多炼钢车间，通过向残余炉渣吹氮，在转炉炉身和耳轴处形成一层保护层，这个过程叫"溅渣护炉"。转炉即将服满炉役时，可能有必要对(转炉炉衬)磨损严重的地方喷补耐火材料。一旦维修完成，转炉又将准备接收下一炉炉料了。

基本的化学平衡和热量平衡

15 以250吨炉容——这跟美国1999年在运转的50座BOF的平均炉容接近——为基础进行下述计算。假设炉料的化学成分如表7.1所示：

表7.1 炉料化学成分表

成分含量%	硅%	锰%	硫%	磷%	铝%	残余成分
铁水	4.5	.75	1.0	.01	.05	0
废钢	.05	.05	.4	.015	.01	.03

16 假设总炉料重275吨,其中75%为铁水,产出钢水250吨(不考虑添加合金元素情况),表7.2给出了每吨铁水的热平衡。如果(炼钢需要的)氧来自于空气,将氮气从室温加热到2900°F(1593.3℃)需要的热量大约为500,000英国热量单位/每吨铁水,显示这里的BOS采用的是贝塞麦工艺,用冷废钢代替冷氮(NTHM指一短吨或2000磅铁水)。

表7.2 炉料中铁水占比75%时每美吨铁水的热平衡

产生热量	Btu(000's)	需要热量	Btu(000's)
C→CO	366	铁水 2400→2900 F	220
Si→SiO_2	204	溶剂→2900 F	110
Mn→MnO	60	O_2→2900 F	120
P→P_2O_5	10	热损	50
Fe→FeO	110	废钢→2900 F	415
CO→CO_2	130		
造渣	35		
总计	915		915

17 炉料中铁水的实际比例对于铁水的硅含量和温度十分敏感,明显随着后两者的降低而增加。

18 每炉需要的氧气如表7.3所示,按"#/美吨铁水"和"不同反应需氧占比"两栏列出。181 #/美吨铁水对应的大约是18.6吨/炉或1800scf/吨出钢。如终点控制系统不佳,需要二次吹氧,则氧气消耗增加。

表7.3 每#/美吨铁水所需氧气

反应	#/美吨铁水	总耗氧占比%
C→CO	120	66
Si→SiO_2	17	9
Fe→FeO(渣)	16	9
CO→CO_2	12	7
Fe→FeO(粉末)	8	4
Mn, P→MnO, P_2O_5	7	4
溶解氧	1	1
	181	100

19 表7.4显示的是收得率损失的最终计算结果。类金属和锰被氧化，离开炉内熔融金属，废钢常被覆盖一层会挥发的锌层，而铁则会进入炉渣、烟气或因喷溅而损失。生产250吨钢水，需要250/0.91也就是275吨炉料，其中206吨是铁水，剩下的是废钢。

表7.4 碱性氧气转炉一个炉次/一次熔炼的收得率损失

损失的物质	占炉量百分比
铁水里的类金属(6.3%)	4.7
废钢上的碎片、涂层(2.5%)	0.6
进入炉渣的铁	2.2
进入烟气的铁	1
喷溅损失的铁	0.5
总计	9

原材料

铁水

20 铁水指高炉生产的饱和含碳高达4.3%、含硅不大于1%的液态铁。铁水通过鱼雷罐车或钢包被运到BOF车间。铁水的化学成分取决于高炉生产方式及装入高炉炉料(含铁材料)的类型。当前的趋势是注重生产率、(尽量)减少炉渣产量及燃料消耗，因此生产出来的铁水含硅较少，含硫较高。如果BOF炉渣被循环利用，则磷和锰的含量急剧上升，因为磷和锰几乎百分之百地回到铁水中。美国产铁矿石含磷和锰都低。

21 高炉(铁水)含硫可达0.05%，但在将铁水用于BOF之前，可以利用高效铁水脱硫设备将硫含量降低到0.01%以下。最常用的脱硫剂有石灰、碳酸钙和镁，可单独或联合使用，通过喷枪喷进铁水。含硫化合物进入炉渣。然而，除非在铁水倒进BOF之前将富含硫的炉渣撇去，否则实际进入转炉的硫还是会远远高于预期。

废钢

22 自热BOS生产时，废钢最吸热。占炉料的20%~25%，废钢也是最重要、最贵的炉料之一。

23 废钢来源多样，最主要的是工厂自身产生的"自产废钢"。随着连铸被应用于生产，自产废钢的量已下降，现在钢铁联合工厂需到市场上去购买废钢。平轧废钢通常质量较高，对BOF生产/化学成分的影响几乎可以忽略。镀锌废钢上有一层镀锌层，所以会有大约2%的收得率损失。"边角料废钢"是在加工生产钢产品过程中产生的，很快就会进入循环利用的过程。很多钢厂都与生产商签有合同，购买他们的边角料废钢。"报废产品废钢"或"用后废钢"在产品使用寿命结束后重回市场。钢罐很快就会重回市场，而汽车的平均寿命却有12年。

24　废钢的大小也不一，化学成分和价格多种多样。所有这些让购买和熔炼废钢十分复杂。很大块的废钢熔炼起来不易，装进转炉后可能会损坏炉体。有些废钢会含油，或表面已氧化。报废产品废钢可能含有很多其他成分，这些成分可能有害或易爆。很明显，这类废钢的化学成分难以准确测定。

25　因为钢制产品种类太多，所以废钢的选择就更加复杂。深冲钢的最大残余（%Cu+%Sn+%Ni+%Cr+%Mo）含量被限制小于0.13%，而其他钢产品允许该数值高至0.80%。由于无法通过氧化从钢中去除这些元素，因此只能通过添加高纯净废钢或铁水将它们在成品中的含量稀释降低。在BOS中使用具有内在稀释功能的低残余铁水，是区分BOF和EAR炼钢的特征之一。

熔剂

26　熔剂的作用主要有两个，其中之一是与铁水中硅氧化后产生的二氧化硅结合生成"碱性"炉渣，炉渣在炼钢温度下处于液态，吸收、除去铁水中的硫和磷。

27　石灰（95+%CaO）和白云石（58%CaO，39%MgO）是两种主要的熔剂。两者均经焙烧碳酸盐矿物而成，焙烧通常是在钢厂外的回转窑中进行的。焙烧$CaCO_3$和$MgCO_3$释放CO_2，得到CaO或MgO。有"软"和"硬"两种类型的烧石灰。一块软烧石灰可迅速溶解于一杯水中，速溶过程中释放热量，而硬烧石灰则什么也不会发生。软烧石灰较硬烧石灰能够更快成渣，这一点对于在短暂的吹氧周期里有效除去硫和磷至关重要。石灰的装炉量取决于铁水中Si的含量。

28　BOS炼钢时，渣中CaO与SiO_2之比较高较理想，如通常的经验是6倍于炉料中Si的重量。MgO的添加量定为最终炉渣重量的8%—10%。这能让渣中的MgO处于饱和状态，从而减少MgO炉衬受到的化学侵蚀。

冷却剂

29　石灰石、废钢和海绵铁都是潜在的冷却剂，可以在炉次过吹、温度过高时加入炉中。至于具体到某一炼钢车间应选择何种冷却剂，则取决于成本和设备上的考虑。

合金料

30　散装合金料从高架料仓装入钢包。常见合金料有锰铁（80%Mn，6%C，balance Fe）、锰硅（66%Mn，16%Si，2%C，balance Fe）和硅铁（75% Si，balance Fe）。铝可以以型

材的形式加入和/或以铝棒的形式插入。硫、碳、钙和像硼、钛一类的特种元素以粉末形式用直径约二分之一英寸的低碳钢套子包起来添加进钢包炉。

耐火材料

31　今天美国用于氧气炼钢炉的大部分耐火砖的主要材料都是氧化镁（MgO），氧化镁可从矿物或海水中获得。美国现在只有一座白云石（MgO+CaO）矿（靠近宾州的雷丁市）在开采。对氧化镁矿来说，硼氧化物含量越低，杂质水平就越低，烧成的耐火砖热强度就越大。碳以沥青（焦油）或石墨的形式加入。

32　去除和吸纳钢水中的磷需要高碱性炉渣制品，选择镁质石灰质耐火材料用作氧气炼钢炉的炉衬，主要是因为这些耐火材料与需要的高度碱性的终渣相容。炼钢时，随着熔池里的硅被氧化并与石灰结合，炉渣碱性在1至4之间，耐火材料直接接触各种碱性不同的炉渣。随着氧吹时间的增加，特别是随着钢中碳含量降到0.2%以下，铁被氧化，熔池中铁氧化物FeO的含量在增加。虽然所有的耐火材料都会被FeO溶解，但MgO会和FeO形成一种固溶体。这意味着，在一定温度范围内，两者以固体形式共存。然而，氧吹后期形成的高浓度的FeO会氧化（耐火）砖里的碳。

33　最初的耐火砖用焦油来黏结/使用焦油做黏结剂，将MgO颗粒涂上一层焦油，然后温压，这代表着BOS工艺的一次巨大进步。回火去除其中的挥发成分。实际使用时，焦油被碳化，颗粒间残余的碳阻止炉渣浸润和FeO的侵蚀。此外，炉体加热过程中，焦油软化，炉衬膨胀的压力减轻。如将纯MgO颗粒砖在高温下烧结并在真空中浸满焦油，则耐火砖的高温强度会增大。然后，由于环保的原因，此类耐火砖不再用于氧气炼钢。

34　今天的工作衬/工作层炉衬用耐火材料主要是以高质量烧结镁砂和高纯度石墨片为原材料、以树脂为黏结剂制成的镁碳砖。树脂黏结的耐火砖无须烧制，含5%~25%的高纯度石墨片及一到多种金属粉末。这种砖需要在350°F—400°F（176.7℃—204.4℃）进行简单的固化处理，以便热固化树脂，让耐火砖的强度更高，因此安装时更易操作。进一步改善耐火砖的措施包括在混合料里使用预熔颗粒。添加少量的金属添加剂（Si、Al和Mg）保护石墨不被氧化，因为这些金属会被优先氧化。使用时，在耐火砖填充空隙受热的一面形成金属碳化物、氮化物和镁铝尖晶石，增加了耐火砖的强度，改善了耐火砖抗渣侵蚀的性能。

35　耐火材料被渣溶解的速度取决于其性能。这些性能与原材料的纯度和晶体粒度以及（耐火砖）的生产工艺直接相关。添加高达15%的高纯度的石墨到氧化镁—碳耐火材料中可以提高其抗腐蚀能力。但超过15%的添加量将会适得其反，因为耐火砖的密度更低了。最终，决定炉衬选择的因素是吨钢用砖和喷补材料的成本以及转炉利用率的需要。

36　炉渣和钢水渗入耐火材料颗粒、液体流动带来的机械磨损以及炉渣的化学腐蚀都会造成炉衬材料的损失。多年来，人们已想出多种办法应对炉衬的磨损。

37 第一，将转炉临界磨损区(防冲击垫区和出钢侧炉衬、倒炉渣线区和耳轴区)衬以最高质量的耐火砖。

38 第二，溅渣护炉。出钢后通过氧枪将高压氮气吹向炉内，将留在炉内的残余液态炉渣喷溅到炉衬上。这一看起来简易的办法能超预期地延长炉衬寿命，从每炉龄几千炉次增加到超过20000炉次。

39 第三，已研制出的仪器可在短时间内测量炉衬表面形状，将使用MgO料浆的喷补效果最大化。

40 第四，在添加的助熔剂里加入白云石(40%MgO)，以生产出含MgO量8%左右的炉渣，这接近炉渣中MgO的饱和含量。

41 第五，改善终点控制，降低FeO的水平，缩短停吹到装料的间隔，减少耐火材料的磨损。

42 然而，如果不能提高钢铁工业所用碱性耐火砖的质量，增加碱性耐火砖的品种，上述办法成效都不会很大。

43 今天的耐火工业正在经历重大的结构性变化，公司不断被并购，北美地区耐火材料供应商的总数下降了很多，很高比例的耐火材料在海外生产，中国成为了最举足轻重的新竞争对手。

环境问题

44 BOS生产带来的环境上的挑战包括：(1)如何捕获并除去转炉排出的灼热、肮脏的一次废气中的污染物；(2)(如何处理)转炉装料和出钢带来的二次排放；(3)如何控制铁水转运、脱硫及钢包冶金操作等辅助操作带来的排放物；(4)如何循环利用和/或处理收集起来的氧化物粉尘或泥浆；(5)如何处置炉渣。

45 在美国，大部分BOF一次炉气处理系统都被设计用于从一次系统的水冷排气罩里生产工厂用蒸汽。大约一半的系统被设计为敞开燃烧，过量空气从排气罩嘴引入，将一氧化碳全部烧完，然后气体被冷却，通过湿式洗涤器或干式静电除尘器净化。在美国剩下的系统为限制燃烧系统，炉气未经燃烧即作处理，经过湿法洗涤净化后，排放前，点火燃烧。无论哪种情形，净化后的气体都必须满足环境保护局制订的颗粒物质(排放)强制标准。

46 限制燃烧系统为能量的回收提供了可能，这在欧洲和日本更为流行。然而在美国，除了生产蒸汽外，没有人试图回收离开转炉废气中的化学能或显热。尽管这意味着浪费了大量的能量(大约70万英国热量单位/吨)，但无论是将敞开式燃烧转换为限制式燃烧系统，还是增加必要的气体收集装置用于限制燃烧系统，收回投资所需时间都在10年以上。此外，要作出这些改变就必须停炉也让这样的想法显得没有实用价值。大部分美国BOF工厂建于20世纪70年代能源危机之前，甚至在今天，美国的能源也较国外相对便宜。

47 BOF装料和出钢时的二次逃逸排放物或氧吹时逃出主排气罩的排放物，可以通过服务于局部排气罩或位于炼钢车间桁架上的高罩蓬排气罩或两种排气罩的排气系统捕获。通常使用纤维除尘器即布袋集尘器收集这些逃逸排放物。同样，进行诸如铁水转运、脱硫或钢包冶金等辅助操作时通常使用局部排气系统将排放物排入纤维过滤器里。

48　一次系统捕获的颗粒物质，不管是从湿法洗涤器出来的泥浆还是静电除尘器收集的干燥粉尘，都必须处理后才能循环利用。来自于湿法洗涤器的泥浆需要额外的干燥处理。与 EAF 粉尘不同的是，BOF 粉尘或泥浆并不是明文规定的有害废物。如果锌含量足够低，这些粉尘或泥浆可以在压块或造球后在高炉或 BOF 炉中循环利用。正在使用的或正在研制的循环利用这种颗粒物质的工艺有很多。

49　BOF 渣通常含 5% 的 MnO 和 1% 的 P_2O_5，常能在高炉中循环利用。钢渣中的石灰吸水，暴露于空气之中后会膨胀，因此其作为填充料的使用受到限制，但人们正在研究将其用于其他商业用途，以最大程度减少需要处理的量。

结语

50　自二战以来，BOS 作为一项关键工艺，改变了美国的钢铁工业。虽然过去没有认识到这一点，但该工艺使得冶炼和连铸能够很好地配合起来，结果是冶炼工艺、精轧质量和收得率提高了几个百分点，每吨产品所需要的粗钢数量下降了很多。

51　BOS 的未来取决于铁水的供应，而后者又取决于焦炭的价格和供应。虽然在装料时减少铁水的量，如小于 70%，BOF 也可以生产，但（那样则必须）添加辅助燃料，结果是生产率降低，成本增加。人们不断地提出替代高炉的工艺，如在南非的萨尔达纳厂（Saldahna Works），BOF-EAF 混合炉已成为现实。但似乎在未来的几十年里，高炉和 BOS 将一直陪伴我们。

五、专题讨论

读者需求与科技翻译

翻译是一种有目的的活动（a purposeful activity），决定任何翻译过程的首要原则是翻译的目的（the prime principle determining any translation process is the purpose）。① 无论译者的目的是什么，都必须通过译文读者的阅读来实现。如果译作完成了却没有任何读者，译者的翻译目的无法达成。因此，满足读者的阅读需求是翻译的应有之义。

---

①　Christiane Nord. *Translating as a Purposeful Activity: Functionalist Approaches Explained* [M]. Shanghai: Shanghai Foreign Language Education Press, 2001: 27.

有经验的译者总是有着较强的读者意识，翻译时充分考虑读者的各种需求，在不改变原文内容的基础上，尽力提升译文的可读性和可接受性。一方面，读者意识要求译者对译文读者负责。另一方面，读者意识又要求译者遵循文本内部的规范，在合适的目的语的框架内，激活原文中的场景，帮助译文读者理解原文中的各种表达。① 具体来说，译者应以译语读者群体的某些共时文化特征为参照，在具体的翻译活动中，通过一系列的决策、选择、行为，满足译语读者群体的某些需要或适应其阅读习惯。② 也就是说，译者在翻译时必须清楚读者是谁、读者有何需求以及如何满足读者需求，从而有针对性地进行翻译。只有这样，译文才能得到读者的认可，翻译的目的方可实现。

一般来说，译作读者的需求大致可分为三类：对信息的需求、对理解译作的需求以及对译作表达规范的需求。这种分类对于科技翻译同样适用。

1. 满足读者对信息的需求

读者阅读科技译作，是为了获取科技信息，否则他们就不会去阅读这类作品了。因此，读者的信息需求理应得到满足。

不同的读者对信息的需求各不相同。有些读者需要原文的全部信息，有些读者只对原文的部分信息感兴趣，还有一些读者感兴趣的可能只是原文的重点信息。可以想见，针对读者不同的信息需求，译者需要采用不同的翻译策略。如果读者需要原文的全部信息，译者就必须采用全译的策略；如果读者需要的只是原文的部分信息或重点信息，那么译者就必须采用变译的策略。

全译很好理解，就是将原文的全部文字都译成目的语。至于变译，则是黄忠廉教授提出的一个概念。黄忠廉给变译下的定义是："变译是译者根据读者的特殊需求采用扩充、取舍、浓缩、阐释、补充、合并、改造等变通手段摄取原作中心内容或部分内容的翻译活动。"③

黄忠廉总共提出了十一种变译方法，分别是：摘译、编译、译述、缩译、综述、述评、译评、改译、阐译、译写、参译。限于篇幅，这里就不对这些变译方法一一展开讨论。感兴趣的读者可以参阅黄忠廉的《翻译变体研究》（中国对外翻译出版公司，2000）或《变译理论》（中国对外翻译出版公司，2002）。这里仅举一例来说明有时采用变译策略的必要性：

本课课文是关于碱性氧气炼钢工艺的，文章涉及碱性氧气炼钢的发展现状、工艺流程、化学平衡和热量平衡、原材料、耐火材料、环境影响等多个方面。如果预期读者对所有这些话题都感兴趣，当然需要全译；但如果预期读者只对碱性氧气炼钢中所用的耐火材

---

① 周晓梅. 中国文学外译中的读者意识问题[J]. 小说译介与传播研究 2018(3).
② 贺文照. 论中国传统译论中的读者观照[J]. 外语与外语教学，2002(6).
③ 黄忠廉. 翻译变体研究[M]. 北京：中国对外翻译出版公司，2000：5.

料感兴趣，全译就是浪费精力了，也增加了读者查找自己所需信息的时间，这种情况下译者只需要摘译 **Refractories** 一节即可。

2. 满足读者对理解译作的需求

译者进行翻译，当然希望读者能够完全理解译作。但有时由于读者知识有限，直接按照原文翻译，读者理解会有困难，这时就需要采取增加词语、添加注释等措施。例如：

There exist several variations on the Basic Oxygen Steelmaking process: top blowing, bottom blowing, and a combination of the two. This study will focus only on the top blowing variation.

碱性氧气转炉炼钢工艺分几种：<u>氧气顶吹</u>、<u>氧气底吹</u>和<u>顶底复合吹氧</u>。本研究将只讨论顶吹炼钢。

原句中，top blowing、bottom blowing、a combination of the two 对应的汉语词语分别是"顶吹""底吹"和"顶吹底吹合并使用"，但这样翻译，专业知识稍缺的读者可能就不太理解，增加词语译成"氧气顶吹""氧气底吹"和"顶底复合吹氧"，读者就容易理解得多。

再来看几个例子：

This rate of production made the process compatible with the continuous casting of slabs, which in turn had an enormous beneficial impact on yields from crude steel to <u>shipped product</u>, and on downstream flat-rolled quality.

这样的生产速度很好地配合了板坯的连铸，而后者又对从粗钢到<u>成品(发运的产品)</u>收得率及下游的平轧质量有着巨大的、积极的影响。

板坯的连铸就是将"粗钢"变成"成品"的过程，如果将 shipped product 直接译为"发运的产品"，不是专业的表达。先译成"成品"，然后附上"(发运的产品)"作为解释，这样表达既规范，又便于读者理解。

Once the hot metal temperature and chemical analysis of the blast furnace hot metal are known, a computer charge models determine the optimum proportions of scrap and hot metal, flux additions, lance height and oxygen blowing time.

一旦知道了铁水的温度和高炉铁水的化学成分，计算机装料模型就会计算出废钢和铁水的最佳配比、熔剂<u>应</u>添加量、氧枪<u>适宜</u>高度和<u>最佳</u>氧吹时间。

相对原句，译句添加了"应""适宜"和"最佳"，在精准表达 optimum 意思的基础上，又顺应了汉语的表达习惯,方便了读者对译句的理解。

  Although it is possible to operate BOFs with reduced hot metal charges, i. e. <70%, there are productivity penalties and costs associated with the supply of auxiliary fuels.
  虽然在装料时减少铁水的量,如小于70%,BOF也可以生产,但那样则必须添加辅助燃料,结果是生产率降低,成本增加。

译文添加了"那样则必须",将隐藏的逻辑关系显性化,同样是为了便于读者理解。

  Some dimensions of a typical 250 ton BOF vessel in the U. S. are: height 34 feet, outside diameter 26 feet, barrel lining thickness 3 feet, and working volume 8000 cubic feet.
  一座美国的250吨BOF转炉的尺寸通常是:高34英尺(约10.4米),外直径26英尺(约7.9米),炉身内衬厚3英尺(约0.9米),工作容积8000立方英尺(约226.5立方米)。

原句用的是英制计量单位"英尺""立方英尺",中国读者难以形成清晰的概念。为了便于汉语读者理解,译者在英制计量单位后添加了相应的公制计量单位"(约10.4米)""(约7.9米)""(约0.9米)""(约226.5立方米)",适应了汉语读者的认知习惯。同样,temperature at turndown can be off by +/-100°F中的"+/-100°F"被译为+/-100°F(37.8℃),既忠实于原文,又让译语读者有亲切感。
  要满足读者对理解译作的需求,译者还需要斟酌语言的使用。如果预期读者是小朋友,就不能使用难度较大的语言;如果预期读者是非专业人士,就应尽量少使用专业术语;而如果预期读者是专家学者,就要注意语言的学术性、逻辑性、严密性,可以适当增加单位译文内信息的含量。比如前面我们曾说过,翻译科技英语时,遇到英语缩写,可以不翻译,直接使用缩写,或者在第一次遇到时,照抄缩写,并将缩写的汉语对应表达括注于缩写之后,下文只用缩写就可以了。不过,这条原则主要适用于译文读者具有一定专业知识的情况。如果译文读者没有足够的专业知识,最好还是每次都将缩写译成汉语,以免读者忘记缩写所指,还要查找资料或前文。
  来看另外一个例子。前面"分析与讲解"中我们说这个句子:

  The magnesia lime type refractories used in lining oxygen steelmaking vessels are selected mainly for their compatibility with the highly basic finishing slags required to remove and retain phosphorus in solution.

可以译成：

去除和吸纳钢水中的磷需要高碱性炉渣制品，选择镁质石灰质耐火材料用作氧气炼钢炉的炉衬，主要是因为这些耐火材料与需要的高度碱性的终渣相容。

也可以译成：

选择镁质石灰质耐火材料用作氧气炼钢炉的炉衬，主要是因为这些耐火材料不易受除去、吸纳钢水中磷需要的高度碱性的终渣的侵蚀。

相较而言，第一种译法更简单，更适合理解能力较差或没有时间仔细阅读的读者；第二种译法小句更长，读者需要更强的理解能力或更长的阅读时间。而如果预期读者是高中学历以下的人士，那么恐怕就要进行更多处理了，比如：

镁质石灰质耐火材料属于强碱性耐火材料，这些耐火材料不会与钢水中的碱性炉渣发生酸碱中和化学反应，不会被碱性炉渣破坏而影响使用寿命，而吸纳、去除钢水中的磷又需要高度碱性的炉渣，所以氧气炼钢炉选择使用镁质石灰质耐火材料，这样就不需要那么频繁地更换耐火材料内衬，造成时间和成本上的损失。

再看一个例子：

Roll forming is a continuous bending operation in which a long strip of metal is passed through consecutive sets of rolls, or stands, each performing only an incremental part of the bend.

句中，each performing only an incremental part of the bend 直译是"每组轧辊只完成弯曲的一个增量部分"，但这样的译法只有专业人士才能看懂。如果想让普通读者看懂，就需要采用非专业的表达法，如"每组轧辊只完成某一方向的弯曲"。

3. 满足读者对表达规范的需求

译作总是以目的语的形式印刷、出版或发布至虚拟空间，自然要遵循目的语中的表达规范。举一个简单的例子：英语中，书名用斜体表示，但译成汉语时，就不能继续用斜体，而应在书名两边加书名号。译作读者总是期待译作符合目的语的表达规范。请看几个例子：

Slag in the ladle results in phosphorus reversion, retarded desulfurization, and possibly "dirty steel".

钢包中的炉渣会让磷重新回到钢水中，延迟脱硫过程，可能带来"不洁钢"。

此处，dirty steel 被反译为"不洁钢"，因为汉语的专业表达中没有"脏钢"的说法。又比如：

Table 1 illustrates the heat balance PER TON OF HOT METAL. It assumes a 75% hot metal in a total charge of 275 tons which yields 250 tons of liquid steel (without alloys).

假设总炉料重 275 吨，其中 75% 为铁水，产出钢水 250 吨(不考虑添加合金元素情况)，表 1 给出了每吨铁水的热平衡。

原文中用大写 PER TON OF HOT METAL 来强调要表达的信息，但汉字没有大小写之分，因此译文对字体进行了加粗，以译语读者熟悉的方式表达强调的意思。此外，汉语先给假设条件，然后陈述假设条件下产生的结果，所以先译后一句，再译前一句。

Furthermore, below 0.2% C, the highly exothermic oxidation of iron takes place to a variable degree along with decarburization.

此外，含碳量低于 0.2% 时，伴随着脱碳过程发生的铁的氧化释放的大量热量会有所变化。

C 是碳的化学元素符号，如果将 below 0.2% C 译为"低于 0.2% 碳时"或"低于 0.2% C 时"，译文就没有可读性，译成"含碳量低于 0.2% 时"读来自然达意，符合汉语的表达规范。

These bricks require a simple curing step at 350 to 400 °F to "thermoset" the resin that makes them very strong and therefore easily handled during installation.

这种砖需要在 350 °F—400 °F（176.7 °C—204.4 °C）进行简单的固化处理，以便热固化树脂，让耐火砖的强度更高，因此安装时更易操作。

350 to 400 °F 被译为"350 °F—400 °F（176.7 °C—204.4 °C）"，在括号中添加对应的摄氏度是为了让中国读者对前面描述的华氏摄氏度有清晰的概念。其中"to"没直译为"至"，用符号"—"表示，是汉语科技文本对这一概念的规范表达，适应我国科技文本的行文规范。

同样原因,Blowing continues for a predetermined time based on the metallic charge chemistry and the melt specification. This is typically 15 to 20 minutes" 中的 15 to 20 minutes 应译成"15—20 分钟"。

Further refinements include using prefused grains in the mix. Small additions of metal additives (Si, Al, and Mg) protect the graphite from oxidation because they are preferentially oxidized.

进一步改善耐火砖的措施包括在混合料里使用预熔颗粒。添加少量的金属添加剂(Si、Al 和 Mg)保护石墨不被氧化,因为这些金属会被优先氧化。

不翻译(Si, Al, and Mg)中的 Si, Al, Mg, 也是考虑了中国读者的阅读习惯,因为 Si, Al, Mg 是国际通用的化学元素符号,中国读者也习惯这种规范表达。采用零翻译的策略翻译这些元素符号,体现了译者的读者意识。另外,汉语中用顿号而不是用逗号表示较短的停顿,所以需要将 Si 和 Al 之间的英语逗号改成汉语顿号。

Figure 2:BOF Vessel in Its Operating Positions. (Ref:Making, Shaping, and Treating of Steel, 11th Edition, Steelmaking and Refining Volume. AISE Steel Foundation, 1998, Pittsburgh PA)

图 2:各种操作状态时的 BOF 炉体。【出处:《炼钢、钢的成形和处理》(第 11 版)"钢的生产和精炼卷",钢铁工程师协会钢基金会(宾夕法尼亚州匹兹堡),1998 年】

译文遵循了中国读者熟悉的参考文献标注规范,体现了译者的读者意识。

总之,科技译者不能简单地满足于准确传达原文的科技信息,而要在充分考虑读者需求的基础上,以读者最能接受同时又满足目的语表达规范的形式,将读者需要的信息展现在读者面前。

◎ 练习七　将下面的句子翻译成汉语。

1. During the main blow period (25 to 80 pct of blow), it remains constant with a value of 2,573 K.

2. In a BOF process the Mn content in the hot metal usually varies from 0.3 to 0.6 wtpct, and the estimated equilibrium value was found to be 1,000 to 10,000 times lower than the actual manganese concentration.

3. The trajectory of a droplet in both vertical and horizontal direction was calculated by the

force balance method with taking into account the dynamic change in density due to bloating caused by the nucleation of CO gas bubbles.

4. The reductant gas from the reformer (75% H_2, 14% CO, 7% CO_2, 4% CH_4) is introduced to the fixed bed reactors in their third cooling step mode.

5. In spite of their early success as the first direct-reduction processes, HyL and HyL II are no longer competitive either in terms of energy consumption, or in terms of operation and investment costs. The process has now been replaced by the more successful HyL III process, which was developed in 1979.

6. The process involves three main components, the shaft reactor where the iron ore is reduced, the gas reformer where the natural gas is reformed to produce CO and H_2, and the reductant gas heater.

7. Once the oxygen was no longer bottom-blown—as in the Thomas or Bessemer methods—but instead top-blown, the oxygen blowing technique became widely popular after the Second World War.

8. Liquid slag will float on the iron and is separated from the iron by the SKIMMER after tapping. Molten iron flows under the skimmer into an iron pool and then over the IRON DAM to the iron runner.

9. Alloy steels are divided into two types, low alloy steels with under 10 per cent of added elements and high alloy steels with over 10 per cent (usually between 15 and 30 per cent) of added elements.

10. Tungsten is used largely with chromium as a kind of high-speed tool steel which contains 14.00 to 18.00 per cent tungsten and 2.00 to 4.00 per cent chromium.

第八课 电 弧 炉

> 相对于钢铁联合企业中以高炉—转炉炼钢为代表的常规流程而言，以废钢为主要原料的电弧炉炼钢生产线具有工序少、投资低和建设周期短的特点，因而被称为短流程。
> ——徐匡迪(中国工程院院士、第十届全国政协副主席、上海市原市长、中国工程院第八届主席团名誉主席)[①]

一、科技术语

electric arc A luminous discharge of current that is formed when a strong current jumps a gap in a circuit or between two electrodes.(电弧)

induction furnace An electrical furnace in which the heat is applied by induction heating of metal.(感性炉)

rendering A drawing in perspective of a proposed structure.(透视图)

electrode A solid electric conductor through which an electric current enters or leaves an electrolytic cell or other medium.(电极)

hydraulic cylinder A mechanical actuator that is used to give a unidirectional force through a unidirectional stroke.(液压缸)

busbar A metallic strip or bar, typically housed inside switchgear, panel boards, and busway enclosures for local high current power distribution.(母线)

dolomite A white or light-colored mineral, essentially $CaMg(CO_3)_2$, used in fertilizer, as a furnace refractory, and as a construction and ceramic material.(白云石)

harmonic distortion Electronics distortion caused by nonlinear characteristics of electronic apparatus, esp. of audio amplifiers, that generate unwanted harmonics of the input frequencies.(谐波失真)

---

① 徐匡迪,洪新. 电炉短流程回顾和发展中的若干问题[J]. 中国冶金, 2005(7): 1.

matte A mixture of a metal with its sulfides, produced by smelting the sulfide ores of copper, lead, or nickel.(锍;熔锍;冰铜、冰铅或冰镍)

affinity The tendency for two substances to combine; chemical attraction.(亲和力)

plasma An electrically neutral, highly ionized phase of matter composed of ions, electrons, and neutral particles.(等离子,等离子体)

二、英语原文

Electric Arc Furnace

1　An electric arc furnace (EAF) is a furnace that heats charged material by means of an electric arc.

2　Industrial arc furnaces range in size from small units of approximately one ton capacity (used in foundries for producing cast iron products) up to about 400 ton units used for secondary steelmaking. Arc furnaces used in research laboratories and by dentists may have a capacity of only a few dozen grams. Industrial electric arc furnace temperatures can be up to 1,800℃ (3,272℉), while laboratory units can exceed 3,000℃ (5,432℉).

3　Arc furnaces differ from induction furnaces in that the charged material is directly exposed to an electric arc, and the current in the furnace terminals passes through the charged material.

History

4　In the 19th century, a number of men had employed an electric arc to melt iron. Sir Humphry Davy conducted an experimental demonstration in 1810; welding was investigated by Pepys in 1815; Pinchon attempted to create an electrothermic furnace in 1853; and, in 1878-1879, Sir William Siemens took out patents for electric furnaces of the arc type.

5　The first electric arc furnaces were developed by Paul Héroult, of France, with a commercial plant established in the United States in 1907. The Sanderson brothers formed The Sanderson Brothers Steel Co. in Syracuse, New York, installing the first electric arc furnace in the U.S. This furnace is now on display at Station Square, Pittsburgh, Pennsylvania.

6　Initially "electric steel" was a specialty product for such uses as machine tools and spring steel. Arc furnaces were also used to prepare calcium carbide for use in carbide lamps. The *Stassano electric furnace* is an arc type furnace that usually rotates to mix the bath. The

Girod furnace is similar to the *Héroult furnace*.

7   While EAFs were widely used in World War II for production of alloy steels, it was only later that electric steelmaking began to expand. The low capital cost for a mini-mill—around US $140 to 200 per ton of annual installed capacity, compared with US $1,000 per ton of annual installed capacity for an integrated steel mill—allowed mills to be quickly established in war-ravaged Europe, and also allowed them to successfully compete with the big United States steelmakers, such as Bethlehem Steel and U.S. Steel, for low-cost, carbon steel "long products" (structural steel, rod and bar, wire, and fasteners) in the U.S. market.

8   When Nucor—now one of the largest steel producers in the U.S.—decided to enter the long products market in 1969, they chose to start up a mini-mill, with an EAF as its steelmaking furnace, soon followed by other manufacturers. Whilst Nucor expanded rapidly in the Eastern US, the companies that followed them into mini-mill operations concentrated on local markets for long products, where the use of an EAF allowed the plants to vary production according to local demand. This pattern was also followed globally, with EAF steel production primarily used for long products, while integrated mills, using blast furnaces and basic oxygen furnaces, cornered the markets for "flat products"—sheet steel and heavier steel plate. In 1987, Nucor made the decision to expand into the flat products market, still using the EAF production method.

Construction

9   An electric arc furnace used for steelmaking consists of a refractory-lined vessel, usually water-cooled in larger sizes, covered with a retractable roof, and through which one or more graphite electrodes enter the furnace. The furnace is primarily split into three sections:
- the shell, which consists of the sidewalls and lower steel "bowl";
- the hearth, which consists of the refractory that lines the lower bowl;
- the roof, which may be refractory-lined or water-cooled, and can be shaped as a section of a sphere, or as a frustum (conical section). The roof also supports the refractory delta in its centre, through which one or more graphite electrodes enter.

10   The hearth may be hemispherical in shape, or in an eccentric bottom tapping furnace, the hearth has the shape of a halved egg. In modern meltshops, the furnace is often raised off the ground floor, so that ladles and slag pots can easily be manufactured under either end of the furnace. Separate from the furnace structure is the electrode support and electrical system, and the tilting platform on which the furnace rests. Two configurations are possible: the electrode supports and the roof tilt with the furnace, or are fixed to the raised platform.

11 A typical alternating current furnace is powered by a three-phase electrical supply and therefore has three electrodes. Electrodes are round in section, and typically in segments with threaded couplings, so that as the electrodes wear, new segments can be added. The roof of an arc furnace removed, showing the three electrodes. The arc forms between the charged material and the electrode, the charge is heated both by current passing through the charge and by the radiant energy evolved by the arc. The electric arc temperature reaches around 3,000℃ (5,000℉), thus causing the lower sections of the electrodes glow incandescently when in operation. The electrodes are automatically raised and lowered by a positioning system, which may use either electric winch hoists or hydraulic cylinders. The regulating system maintains approximately constant current and power input during the melting of the charge, even though scrap may move under the electrodes as it melts. The mast arms holding the electrodes can either carry heavy busbars (which may be hollow water-cooled copper pipes carrying current to the electrode clamps) or be "hot arms", where the whole arm carries the current, increasing efficiency. Hot arms can be made from copper-clad steel or aluminum. Since the electrodes move up and down automatically for regulation of the arc, and are raised to allow removal of the furnace roof, large water-cooled cables connect the bus tubes/arms with the transformer located adjacent to the furnace. To protect the transformer from heat, it is installed in a vault and is itself cooled via pumped oil exchanging heat with the plant's water-cooling systems, as the electrical conditions for arc-furnace steelmaking are extremely stressful on the transformer.

12 The furnace is built on a tilting platform so that the liquid steel can be poured into another vessel for transport. The operation of tilting the furnace to pour molten steel is called "tapping". Originally, all steelmaking furnaces had a tapping spout closed with refractory that washed out when the furnace was tilted, but often modern furnaces have an eccentric bottom tap-hole (EBT) to reduce inclusion of nitrogen and slag in the liquid steel. These furnaces have a taphole that passes vertically through the hearth and shell, and is set off-centre in the narrow "nose" of the egg-shaped hearth. It is filled with refractory sand, such as olivine, when it is closed off. Modern plants may have two shells with a single set of electrodes that can be transferred between the two; one shell preheats scrap while the other shell is utilized for meltdown. Other DC-based furnaces have a similar arrangement, but have electrodes for each shell and one set of electronics.

13 AC furnaces usually exhibit a pattern of hot and cold-spots around the hearth perimeter, with the cold-spots located between the electrodes. Modern furnaces mount oxygen-fuel burners in the sidewall and use them to provide chemical energy to the cold-spots, making the heating of the steel more uniform. Additional chemical energy is provided by injecting

oxygen and carbon into the furnace; historically this was done through lances (hollow mild-steel tubes) in the slag door, now this is mainly done through multiple wall-mounted injection units that combine the oxygen-fuel burners and the oxygen or carbon injection systems into one unit.

14　A mid-sized modern steelmaking furnace would have a transformer rated about 60,000,000 volt-amperes (60 MVA), with a secondary voltage between 400 and 900 volts and a secondary current in excess of 44,000 amperes. In a modern shop such a furnace would be expected to produce a quantity of 80 metric tonnes of liquid steel in approximately 50 minutes from charging with cold scrap to tapping the furnace. In comparison, basic oxygen furnaces can have a capacity of 150 to 300 tonnes per batch, or "heat", and can produce a heat in 30 to 40 minutes. Enormous variations exist in furnace design details and operation, depending on the end product and local conditions, as well as ongoing research to improve furnace efficiency. The largest scrap-only furnace (in terms of tapping weight and transformer rating) is a DC furnace operated by Tokyo Steel in Japan, with a tap weight of 420 metric tonnes and fed by eight 32 MVA transformers for 256 MVA total power.

15　To produce a ton of steel in an electric arc furnace requires approximately 400 kilowatt-hours per short ton or about 440 kWh per metric tonne; the theoretical minimum amount of energy required to melt a tonne of scrap steel is 300 kWh (melting point 1,520℃/ 2,768℉). Therefore, a 300-tonne, 300 MVA EAF will require approximately 132 MWh of energy to melt the steel, and a "power-on time" (the time that steel is being melted with an arc) of approximately 37 minutes. Electric arc steelmaking is only economical where there is plentiful electricity, with a well-developed electrical grid. In many locations, mills operate during off-peak hours when utilities have surplus power generating capacity and the price of electricity is less.

Operation

16　Scrap metal is delivered to a scrap bay, located next to the melt shop. Scrap generally comes in two main grades: shred (white goods, cars and other objects made of similar light-gauge steel) and heavy melt (large slabs and beams), along with some direct reduced iron (DRI) or pig iron for chemical balance. Some furnaces melt almost 100% DRI.

17　The scrap is loaded into large buckets called baskets, with "clamshell" doors for a base. Care is taken to layer the scrap in the basket to ensure good furnace operation; heavy melt is placed on top of a light layer of protective shred, on top of which is placed more shred. These layers should be present in the furnace after charging. After loading, the basket may pass to a scrap pre-heater, which uses hot furnace off-gases to heat the scrap and recover energy, increasing plant efficiency.

18   The scrap basket is then taken to the melt shop, the roof is swung off the furnace, and the furnace is charged with scrap from the basket. Charging is one of the more dangerous operations for the EAF operators. A lot of potential energy is released by multiple tonnes of falling metal; any liquid metal in the furnace is often displaced upwards and outwards by the solid scrap, and the grease and dust on the scrap is ignited if the furnace is hot, resulting in a fireball erupting. In some twin-shell furnaces, the scrap is charged into the second shell while the first is being melted down, and pre-heated with off-gas from the active shell. Other operations are continuous charging—pre-heating scrap on a conveyor belt, which then discharges the scrap into the furnace proper, or charging the scrap from a shaft set above the furnace, with off-gases directed through the shaft. Other furnaces can be charged with hot (molten) metal from other operations.

19   After charging, the roof is swung back over the furnace and meltdown commences. The electrodes are lowered onto the scrap, an arc is struck and the electrodes are then set to bore into the layer of shred at the top of the furnace. Lower voltages are selected for this first part of the operation to protect the roof and walls from excessive heat and damage from the arcs. Once the electrodes have reached the heavy melt at the base of the furnace and the arcs are shielded by the scrap, the voltage can be increased and the electrodes raised slightly, lengthening the arcs and increasing power to the melt. This enables a molten pool to form more rapidly, reducing tap-to-tap times. Oxygen is blown into the scrap, combusting or cutting the steel, and extra chemical heat is provided by wall-mounted oxygen-fuel burners. Both processes accelerate scrap meltdown. Supersonic nozzles enable oxygen jets to penetrate foaming slag and reach the liquid bath.

20   An important part of steelmaking is the formation of slag, which floats on the surface of the molten steel. Slag usually consists of metal oxides, and acts as a destination for oxidized impurities, as a thermal blanket (stopping excessive heat loss) and helping to reduce erosion of the refractory lining. For a furnace with basic refractories, which includes most carbon steel-producing furnaces, the usual slag formers are calcium oxide (CaO, in the form of burnt lime) and magnesium oxide (MgO, in the form of dolomite and magnesite). These slag formers are either charged with the scrap, or blown into the furnace during meltdown. Another major component of EAF slag is iron oxide from steel combusting with the injected oxygen. Later in the heat, carbon (in the form of coke or coal) is injected into this slag layer, reacting with the iron oxide to form metallic iron and carbon monoxide gas, which then causes the slag to foam, allowing greater thermal efficiency, and better arc stability and electrical efficiency. The slag blanket also covers the arcs, preventing damage to the furnace roof and sidewalls from radiant

heat.

21 Once the scrap has completely melted down and a flat bath is reached, another bucket of scrap can be charged into the furnace and melted down, although EAF development is moving towards single-charge designs. After the second charge is completely melted, refining operations take place to check and correct the steel chemistry and superheat the melt above its freezing temperature in preparation for tapping. More slag formers are introduced and more oxygen is blown into the bath, burning out impurities such as silicon, sulfur, phosphorus, aluminum, manganese, and calcium, and removing their oxides to the slag. Removal of carbon takes place after these elements have burnt out first, as they have a greater affinity for oxygen. Metals that have a poorer affinity for oxygen than iron, such as nickel and copper, cannot be removed through oxidation and must be controlled through scrap chemistry alone, such as introducing the direct reduced iron and pig iron mentioned earlier. A foaming slag is maintained throughout, and often overflows the furnace to pour out of the slag door into the slag pit. Temperature sampling and chemical sampling take place via automatic lances. Oxygen and carbon can be automatically measured via special probes that dip into the steel, but for all other elements, a "chill" sample—a small, solidified sample of the steel—is analyzed on an arc-emission spectrometer.

22 Once the temperature and chemistry are correct, the steel is tapped out into a preheated ladle through tilting the furnace. For plain-carbon steel furnaces, as soon as slag is detected during tapping the furnace is rapidly tilted back towards the deslagging side, minimizing slag carryover into the ladle. For some special steel grades, including stainless steel, the slag is poured into the ladle as well, to be treated at the ladle furnace to recover valuable alloying elements. During tapping some alloy additions are introduced into the metal stream, and more lime is added on top of the ladle to begin building a new slag layer. Often, a few tonnes of liquid steel and slag are left in the furnace in order to form a "hot heel", which helps preheat the next charge of scrap and accelerate its meltdown. During and after tapping, the furnace is "turned around": the slag door is cleaned of solidified slag, the visible refractories are inspected and water-cooled components checked for leaks, and electrodes are inspected for damage or lengthened through the addition of new segments; the taphole is filled with sand at the completion of tapping. For a 90-tonne, medium-power furnace, the whole process will usually take about 60-70 minutes from the tapping of one heat to the tapping of the next (the tap-to-tap time).

23 The furnace is completely emptied of steel and slag on a regular basis so that an inspection of the refractories can be made and larger repairs made if necessary. As the

refractories are often made from calcined carbonates, they are extremely susceptible to hydration from water, so any suspected leaks from water-cooled components are treated extremely seriously, beyond the immediate concern of potential steam explosions. Excessive refractory wear can lead to breakouts, where the liquid metal and slag penetrate the refractory and furnace shell and escape into the surrounding areas.

Advantages of Electric Arc Furnace for Steelmaking

24 The use of EAFs allows steel to be made from a 100% scrap metal feedstock. This greatly reduces the energy required to make steel when compared with primary steelmaking from ores. Another benefit is flexibility: while blast furnaces cannot vary their production by much and can remain in operation for years at a time, EAFs can be rapidly started and stopped, allowing the steel mill to vary production according to demand. Although steelmaking arc furnaces generally use scrap steel as their primary feedstock, if hot metal from a blast furnace or direct-reduced iron is available economically, these can also be used as furnace feed. As EAFs require large quantities of electrical power, many companies schedule their operations to take advantage of off peak electricity pricing.

25 A typical steelmaking arc furnace is the source of steel for a mini-mill, which may make bars or strip product. Mini-mills can be sited relatively near to the markets for steel products, and the transport requirements are less than for an integrated mill, which would commonly be sited near a harbour for access to shipping.

Environmental Issues

26 Although the modern electric arc furnace is a highly efficient recycler of steel scrap, operation of an arc furnace shop can have adverse environmental effects. Much of the capital cost of a new installation will be devoted to systems intended to reduce these effects, which include:

Enclosures to reduce high sound levels

Dust collector for furnace off-gas

Slag production

Cooling water demand

Heavy truck traffic for scrap, materials handling, and product

Environmental effects of electricity generation

27 Because of the very dynamic quality of the arc furnace load, power systems may require technical measures to maintain the quality of power for other customers; flicker and harmonic distortion are common side-effects of arc furnace operation on a power system. For this

reason the power station should be located as close to the EA furnaces as possible.

Other Electric Arc Furnaces

28 For steelmaking, direct current (DC) arc furnaces are used, with a single electrode in the roof and the current return through a conductive bottom lining or conductive pins in the base. The advantage of DC is lower electrode consumption per ton of steel produced, since only one electrode is used, as well as less electrical harmonics and other similar problems. The size of DC arc furnaces is limited by the current carrying capacity of available electrodes, and the maximum allowable voltage. Maintenance of the conductive furnace hearth is a bottleneck in extended operation of a DC arc furnace.

29 In a steel plant, a ladle furnace (LF) is used to maintain the temperature of liquid steel during processing after tapping from EAF or to change the alloy composition. The ladle is used for the first purpose when there is a delay later in the steelmaking process. The ladle furnace consists of a refractory roof, a heating system, and, when applicable, a provision for injecting argon gas into the bottom of the melt for stirring. Unlike a scrap melting furnace, a ladle furnace does not have a tilting or scrap charging mechanism.

30 Electric arc furnaces are also used for production of calcium carbide, ferroalloys and other non-ferrous alloys, and for production of phosphorus. Furnaces for these services are physically different from steel-making furnaces and may operate on a continuous, rather than batch, basis. Continuous process furnaces may also use paste-type, Søderberg electrodes to prevent interruptions due to electrode changes. Such a furnace is known as a submerged arc furnace because the electrode tips are buried in the slag/charge, and arcing occurs through the slag, between the matte and the electrode. A steelmaking arc furnace, by comparison, arcs in the open. The key is the electrical resistance, which is what generates the heat required: the resistance in a steelmaking furnace is the atmosphere, while in a submerged-arc furnace the slag or charge forms the resistance. The liquid metal formed in either furnace is too conductive to form an effective heat-generating resistance.

31 Amateurs have constructed a variety of arc furnaces, often based on electric arc welding kits contained by silical blocks or flower pots. Though crude, these simple furnaces can melt a wide range of materials, create calcium carbide, etc.

Electric Arc Furnace Cooling Methods

32 Non-pressurized water-cooling. This process also called spray cooling performs its cooling function at atmospheric pressure utilizing spray nozzles to supply cooling water to roofs,

sidewalls and off gas ductwork on electric arc furnaces (EAF). Each nozzle is located and sized to provide the required amount of cooling water for the varying heat loads through the equipment. Spray nozzles provide uniform turbulence across the hot face providing an efficient cooling method. This technology was developed as a safer alternative by steel makers and patented by Union Carbide Corporation. Systems Spray-Cooled Inc. later purchased this patented cooling technology, a member of The Systems Group.

Plasma Arc Furnace

33 A plasma arc furnace (PAF) uses plasma torches instead of graphite electrodes. Each of these torches consists of a casing provided with a nozzle and an axial tubing for feeding a plasma-forming gas (either nitrogen or argon), and a burnable cylindrical graphite electrode located within the tubing. Such furnaces can be referred to as "PAM" (Plasma Arc Melt) furnaces. They are used extensively in the titanium melt industry and similar specialty metals industries.

Vacuum Arc Remelting

34 Vacuum arc remelting (VAR) is a secondary remelting process for vacuum refining and manufacturing of ingots with improved chemical and mechanical homogeneity.

35 In critical military and commercial aerospace applications, material engineers commonly specify VIM-VAR steels. VIM means Vacuum Induction Melted and VAR means Vacuum Arc Remelted. VIM-VAR steels become bearings for jet engines, rotor shafts for military helicopters, flap actuators for fighter jets, gears in jet or helicopter transmissions, mounts or fasteners for jet engines, jet tail hooks and other demanding applications. Most grades of steel are melted once and are then cast or teemed into a solid form prior to extensive forging or rolling to a metallurgically sound form. In contrast, VIM-VAR steels go through two more highly purifying melts under vacuum. After melting in an electric arc furnace and alloying in an argon oxygen decarburization vessel, steels destined for vacuum remelting are cast into ingot molds. The solidified ingots then head for a vacuum induction melting furnace. This vacuum remelting process rids the steel of inclusions and unwanted gases while optimizing the chemical composition. The VIM operation returns these solid ingots to the molten state in the contaminant-free void of a vacuum. This tightly controlled melt often requires up to 24 hours. Still enveloped by the vacuum, the hot metal flows from the VIM furnace crucible into giant electrode molds. A typical electrode stands about 15 feet (5 m) tall and will be in various diameters. The electrodes solidify under vacuum.

36 For VIM-VAR steels, the surface of the cooled electrodes must be ground to remove surface irregularities and impurities before the next vacuum remelt. Then the ground electrode is placed in a VAR furnace. In a VAR furnace the steel gradually melts drop-by-drop in the vacuum-sealed chamber. Vacuum arc remelting further removes lingering inclusions to provide superior steel cleanliness and further remove gases such as oxygen, nitrogen and hydrogen. Controlling the rate at which these droplets form and solidify ensures a consistency of chemistry and microstructure throughout the entire VIM-VAR ingot. This in turn makes the steel more resistant to fracture or fatigue. This refinement process is essential to meet the performance characteristics of parts like a helicopter rotor shaft, a flap actuator on a military jet or a bearing in a jet engine.

37 For some commercial or military applications, steel alloys may go through only one vacuum remelt, namely the VAR. For example, steels for solid rocket cases, landing gears or torsion bars for fighting vehicles typically involve the one vacuum remelt.

38 Vacuum arc remelting is also used in production of titanium and other metals which are reactive or in which high purity is required.

三、分析与讲解

电力的发现和应用不仅改变了人类的生活方式，也改变了人类的生产方式。作为动力源的一种，电力被广泛地应用在各种各样的工厂里替代人力、热力等传统动力，极大地改善了工厂的生产条件，促进了人们生活水平的提高。

既然电蕴含着巨大的能量，将电力用于钢铁生产也就是顺理成章的事了。电力应用于钢铁生产最典型的例子便是电弧炉炼钢。如同本篇文章所叙述的，用电弧炼铁起始于19世纪，而第一座电弧炉是由法国的保罗·埃鲁研制的。

在我国，最早用电弧炉冶炼合金钢及铁合金的，是中国金属学会首任理事长、前中科院上海冶金所原所长、已故的周仁院士。他于抗战最艰苦的1941年在昆明西山脚下独立自主设计、建设了中国电气制钢厂，用一台小电弧炉为战时中国的军械修理厂提供合金钢刀具和汽车修配用钢。①

电弧炉炼钢具有流程短、绿色环保等优点，代表着钢铁行业的发展趋势。在美国、欧盟等地区，短流程电弧炉炼钢占比已超过钢产量的40%，而我国钢铁生产仍以铁矿石为原料的长流程炼钢为主，占比高达90%以上，两者差距十分明显，值得所有钢铁行业人士反思。

从语言上来说，这篇文章总的来说较为简洁，翻译时需要注意的细节不多。各个段落

---

① 徐匡迪. 中国特钢生产60年[J]. 钢铁，2014(7)：3.

中值得关注的地方见如下分析：

第 1 段：原句关于 EAF 的定义，汉语定义表达需简洁紧凑，需要将定语从句译成前置修饰语。

第 3 段：原句中 the charged material 后事实上省略了 in arc furnaces，翻译时需要体现出来。

第 3、4 段间小标题：原标题是个简单的 History，但如果直译成"历史"，有点意犹未尽，最好译成"历史发展"。

第 4 段：(1)段中有四个表示时间的介词短语，按中文习惯，表达时需将四个时间词语放在小句前，形成排比，后面的句子最好采用同样的句式。(2)took out 在中文中没有对应的词语，可以译为"申请获批"。

第 5 段：需要注意的是，第一句中的 in 1907 修饰的是 established，而不是 were developed。

第 7 段：annual installed capacity 字面意思是"每年已安装能力"，这里指"年产能"。

第 8 段：This pattern was also followed globally 是个被动句，翻译时可以译成汉语被动句"这种模式也在全球被模仿"，但主动句"这种模式也在全球传播开来"更符合阅读习惯。

第 9 段：(1)翻译这段时需要注意句子之间的连贯，最好将 usually water-cooled in larger sizes 提出来，放在 through which one or more graphite electrodes enter the furnace 译句的后面。(2)lower steel "bowl" 中的 bowl 是个比喻，指炉壳下面呈碗形，可译成"下面碗形部分的钢制外壳"。

第 11 段：(1)typically in segments with threaded couplings 翻译起来有点难度，这里指"(电极)通常分几段，用线连接起来"，方便电极磨损后，更换电极段。如果电极是一个整体，部分地方磨损就更换，会造成资源浪费。(2)large water-cooled cables 中的 large 修饰的是 cables，指电缆截面大，或者说电缆较粗。(3)To protect the transformer from heat 需要意译成"为了保护变压器不受高温的影响"，不能直译 heat。

第 12 段：(1)that washed out 虽然使用的是主动语态，但有被动的意思，不妨译成"耐火材料被冲走"。(2)Modern plants may have two shells 中的 plants 最好译成"车间"而不是"工厂"，因为一个工厂可能会有多个车间，而每个车间都可以装有两座电弧炉。

第 17 段：Care is taken to layer the scrap in the basket 中的 layer 是"一层层放"的意思，这里可以译为"铺放"或者"铺"。

第 18 段：Other operations are continuous charging——pre-heating scrap on a conveyor belt 中的 operations 虽然字面意思是"操作"，但这里需要译成"电弧炉"，而 Other furnaces can be charged with hot (molten) metal from other operations 中的 operations 需要译成"设备"：还有些电弧炉从别的设备里直接装入铁水(熔融金属)。

第 20 段：(1)Later in the heat 中的 heat 指"一炉钢"，所以这个词组的意思是"在冶炼

163

一炉钢的后期",也就是"冶炼后期"。(2) greater thermal efficiency, and better arc stability and electrical efficiency 事实上是三个平行词组 greater thermal efficiency, better arc stability, better electrical efficiency,所以翻译时应使用同样的结构来表达。

第 27 段：the very dynamic quality of the arc furnace load 是一种抽象的说法，具体说来就是"电弧炉的负载变化非常大"。

第 30 段：A steelmaking arc furnace, by comparison, arcs in the open 这句中 arcs in the open 如果直译成"在空气中起弧"也可以，但不如译为"起弧发生在炉料之外"，因为前者将无关的信息"空气"带入了译文。

第 35 段：VIM-VAR steels 可直接翻译成"VIM-VAR 钢"；后面的 VIM 和 VAR 可以不翻译，直接使用缩写；Vacuum Induction Melted 和 Vacuum Arc Remelted 分别是 VIM、VAR 的全称，说明 VIM 和 VAR 是怎么来的，处理时应该先照抄原文然后用括号带上两个术语相应的汉语译文。

汉语科技文献中也大量使用英文缩写，因此英语原文中的缩写很多时候可以不翻译，或者在第一次遇到时在缩写之后用括号带出缩写的意思，下文只用缩写，以节省笔墨。

四、参考译文

电 弧 炉

1 电弧炉指利用电弧加热所装物料的炉子。

2 工业用电弧炉大小不一，小电弧炉的炉容大约 1 吨(用于铸造厂生产铸铁产品)，大电弧炉炉容高达 400 吨，用于炉外精炼。用于实验室研究的电弧炉和牙科医生使用的电弧炉的容积可能只有几十克。工业用电弧炉内的温度可以达到 1800℃ (3272℉)，而实验室用电弧炉内的温度能够超过 3000℃ (5432℉)。

3 电弧炉不同于感应炉之处在于，(在电弧炉中)炉料直接暴露于电弧之下，炉内电极间的电流穿过炉料。

4 19 世纪，不少人都使用电弧来熔化铁。1810 年，汉弗莱·戴维爵士进行了实验演示；1815 年，佩匹斯研究了(电弧)焊接；1853 年，帕松尝试研制电热炉；1978 年至 1879 年，威廉·西门子爵士申请获批电弧型电炉。

5 第一座电弧炉是法国人保罗·埃鲁/赫劳尔特研制的。1907 年，一座商用电弧炉冶炼厂在美国建成。桑德森兄弟在纽约州的雪城成立了桑德森兄弟钢铁公司，安装了美国第一座电弧炉。该炉现在宾夕法尼亚州的匹兹堡市车站广场展出。

6 开始时，"电(弧炉)钢"(仅)用于特殊产品，如机床和弹簧钢。电弧炉也用来生产电石，用在电石灯上。斯塔桑诺电炉属电弧型炉，常旋转以混合熔池(里的物料)。吉罗德电炉与埃鲁/赫劳尔特电弧炉类似。

7　尽管电弧炉在"二战"期间就已广泛用于合金钢的生产，电弧炉炼钢的蓬勃发展却是后来的事。小型(电弧炉)钢厂投资成本较低，年产能每吨 140—200 美元。与之相比，投资一体化钢厂，年产能每吨需要 1000 美元。这让(电弧炉)钢厂在饱受战争蹂躏的欧洲迅速发展起来，并能在美国市场与美国大型炼钢厂如伯利恒钢铁公司和美国钢铁公司在低成本碳钢长材(结构钢、线棒材、钢丝和紧固件)上成功竞争。

8　当现今美国最大的钢铁生产商之一的纽科钢铁公司在 1969 年决定进入长材市场时，他们选择开办一家小钢厂，用电弧炉做炼钢炉。这一做法很快为别的钢铁生产商效仿。纽科在美国东部迅速发展起来，效仿开展小钢厂生产的其他公司集中关注的是本地的长材市场，电弧炉的使用能让工厂根据本地需求调整生产/产品。这种模式在全球传播开来，形成电弧炉炼钢主要生产长材而采用高炉和碱性氧吹转炉炼钢的一体化钢厂垄断扁平材——薄板和厚板——市场的局面。1987 年，纽科决定扩大生产进入扁平材市场，仍然采用电弧炉生产工艺。

电弧炉的结构

9　用于炼钢的电弧炉包括一个内衬耐火材料的炉身，炉身之上覆有可伸缩炉顶，通过炉顶，一个或多个石墨电极可进入炉内。大型电弧炉的炉身常用水冷方式冷却。电弧炉主要分成三个部分：

1)炉壳，包括炉壁和下面碗形部分的钢制外壳。

2)炉缸，即用作下面碗形部分内衬的耐火材料。

3)炉顶，内衬耐火砖或水冷冷却，形状可以类似球体的一部分或平截头体(圆锥的一部分)。炉顶同时支撑中央的耐火三角形区域，通过该区域，一个或多个电极伸入炉内。

10　炉缸可以是半球形状，或者在偏心炉底出钢炉中，炉缸的形状像半个鸡蛋。在现代化冶炼车间，电弧炉常被抬高离开地面，以便在电弧炉的任一端下方可方便地制造钢水包和渣罐。跟电弧炉结构分开的是电极支架和电器系统以及电弧炉置于之上的倾炉平台。设计起来有两种选择：电极支架和炉顶与炉身一起倾斜或电极支架和炉顶固定在抬起的平台上。

11　典型的交流电弧炉由三相电源供电，因此有三个电极。电极截面呈圆形，通常分几段，用线连接起来，这样电极磨损后，能够增加新的(电极)段。电弧在所装炉料和电极之间形成，炉料同时被穿过炉料的电流和电弧释放出的辐射能加热。电弧温度达到 3000℃ (5000°F)左右，因此生产时电极下部会发出炽热的光。电极由定位系统自动抬起或降下，定位系统可能使用电动卷扬提升机或液压油缸。调节系统在炉料熔化过程保持电流和功率几乎不变，即使废钢熔化时废钢可能会在电极下移动。支撑电极的横臂可能携带有沉重的母线(母线可能是中空的水冷铜管，将电流输送到电极夹持器)，或者横臂就是"热臂"，整个横臂输送电流，提高效率。热臂可以是包铜的钢或铝。因为电极自动上下运动调节电弧，需要移去炉顶时抬起电极，所以采用了大(截面)水冷电缆，联接母线管/臂和电炉附近的变压器。为了保护变压器不受高温的影响，变压器被安装在带拱顶的结构里，通过泵送的燃油与钢厂的水冷系统交换热量而冷却。这是因为电弧炉炼钢车间的电力环境对于变压器来说是个巨大的考验。

12　电弧炉安装在倾炉平台上，以便钢水能够倾倒到另一容器进行运输。将电弧炉倾斜倾倒钢水的操作被称为"出钢"。最初，所有的炼钢炉都有一个出钢槽，出钢槽为耐火材料封住，炼钢炉倾斜时，耐火材料被冲走，但现代炼钢炉大多配有一个偏心炉底出钢口，以减少钢水中的氮和渣成分。这些炼钢炉有一个垂直穿过炉缸和炉壳的出钢口，位置在蛋形炉缸的狭窄的"鼻子"上，偏离中心。出钢口关闭时装满耐火砂，如橄榄岩。现代电弧炉炼钢车间可能会有两座电弧炉，但只有一组电极，电极在两座炉间移动；一座炉子用于预热废钢，另一座炉子用于冶炼。其他直流电弧炉(车间)的布局类似，但每一座炉子都有电极，(不过)电子系统只有一套。

13　沿着直流电弧炉的炉缸周长通常会分布有热斑和冷斑，冷斑位于电极之间。现代化电弧炉在炉壁装有氧燃烧嘴，通过氧燃烧嘴给冷斑处提供化学能，使得钢的加热更均匀。(此外还)通过向炉中喷吹氧和碳提供额外的化学能。过去喷吹氧和碳是经由渣门通过喷枪(中空低碳钢管)实现的，今天，则主要通过将氧-燃烧嘴和氧/碳喷吹系统合二为一安装在炉壁上的多孔喷吹装置来实现。

14　一座中型的现代化炼钢用电弧炉需要配备一个额定功率大约6000万伏安(60兆伏安)的变压器，二次电压在400—900伏特之间，二次电流超过44000安培。一座现代化电弧炉在从装入废钢到出钢的大约50分钟内预期可生产80公吨钢水。相比之下，碱性氧气转炉每炉产钢150—300公吨，每30~40分钟就可以炼好一炉钢。取决于终端产品、当地情况以及正在开展的提高电弧炉生产效率的研究，电弧炉的设计细节和操作有着巨大的区别。(世界上)最大的(就出钢重量和变压器额定功率而言)只以废钢为原料的电弧炉是日本东京制铁株式会社运营的直流电弧炉，出钢量达到420公吨，由8台32MVA变压器供电，总功率达256MVA。

15　电弧炉生产一吨钢约需400千瓦时/短吨或440千瓦时/公吨。理论上，熔化一吨废钢最少需要300千瓦时的电力(熔点1520℃/2768°F)因此，一座炉容300吨、300MVA的电弧炉需要大约132MWh的电力来熔化废钢，带电时间(用电弧熔化废钢的时间)接近37分钟。电弧炉炼钢只有在电力充足，电网发达的地区才经济。在很多地方，电弧炉只在电力公司电力富余、电价较低的用电低谷时段才生产。

电弧炉的操作

16　将废金属运至冶炼车间相邻的废钢跨。废金属通常有两种形式：碎废钢(家用电器、小汽车和其他由类似轻型钢材制成的物件)和重熔废钢(大型板坯和钢梁)。此外，为了电弧炉内的化学平衡，也会加入一些直接还原铁或生铁。有些电弧炉几乎百分之百使用直接还原铁。

17　将废金属装进被称为料筐的大料斗，料筐底部是蛤形门。将废金属铺在料筐时要小心翼翼，以确保电弧炉的良好运转。重熔废金属须放在保护性的碎金属薄层之上，重熔废金属上面还要放上更多的碎金属。装料入炉后，这些层应保持原状。料筐装满后，移至废金属预热器，预热器使用热炉废气加热废金属，回收热量，提高生产效率。

18　料筐然后被运至冶炼车间。电弧炉的炉顶被移开，从料筐向炉内装料。对于电弧炉操作工来说，装料操作是比较危险的操作之一。许多吨的金属落下会释放大量的势能。固体废金属将炉内可能有的液态金属向上、向外挤出。如果炉内温度高的话，废金属上的油脂和灰尘会被点燃，喷出火球。在一些双炉壳电弧炉里，第一座电弧炉冶炼的同时，废金属被装进第二座电弧炉，通过第一座电弧炉排出的炉气预热。另外一些电弧炉采用连续装料，在皮带运输机上预热废金属，然后皮带运输机将废金属装进炉内，或者通过炉子上方的垂直通道装料，废气也通过垂直通道逸出。还有些电弧炉从别的设备里直接装入铁水(熔融金属)。

19　装好料后，将炉顶移回炉子上，冶炼开始。(首先)，下降电极到废金属之上，产生电弧，然后将电极埋进炉子上部的碎废钢层。冶炼开始阶段选用较低的电压，以保护炉顶和炉壁不受过热或电弧的损坏。一旦电极抵达炉底的重熔金属，因为废金属包住了电弧，所以可以提高电压，稍微抬高电极，增加电弧的长度，加大冶炼功率。这能让熔池更快形成，减少冶炼时间。(同时)，将氧气吹进废金属，燃烧或切割废钢，安装在炉壁上的氧燃烧嘴提供额外的化学能。两者均能提高冶炼速度。超音速喷嘴能让氧气射流穿过泡沫渣，抵达熔池。

20　炼钢过程中一个重要的环节是成渣。炉渣浮在钢水的表面。炉渣通常由金属氧化物构成，是杂质成分氧化后的目的地，起着隔热碳的作用，阻止过量的热损失，帮助降低耐火炉衬的侵蚀。碱性耐火材料电弧炉——这包括大多数生产碳钢的炉子，常用氧化钙(CaO，以煅石灰的形式)或氧化镁(MgO，以白云石和菱镁矿的形式)作成渣剂。这些成渣剂或者与废金属一起装炉，或者在冶炼过程中喷吹进炉。电弧炉炉渣的另一主要成分是与喷吹进的氧一起燃烧的钢形成的铁氧化物。冶炼后期，碳以焦炭或煤的形式被喷吹进渣层，与铁氧化物发生反应，生产金属铁或一氧化碳炉气。一氧化碳炉气让炉渣起泡，能够让热效率更高，电弧更稳定，电力利用更充分。炉渣形成的"毯子"盖住了电弧，防止辐射热损坏炉顶和炉壁。

21　一旦废金属完全熔化，熔池表面变平，就可以加入另一料筐废金属料熔化，虽然电弧炉的发展趋势是设计为只装料一次。第二次装的料完全熔化后，精炼操作开始，目的是检查、调整钢的化学成分，同时将钢水加热到凝固温度之上，准备出钢。(精炼时)，向熔池加入更多的造渣剂，喷吹进更多的氧气，烧尽像硅、硫、磷、铝、锰和钙等杂质，将这类杂质的氧化物移入炉渣。碳的去除发生在这些杂质被烧尽之后，因为这些杂质跟氧的亲和力更大。跟氧的亲和力较铁弱的金属如镍、铜等无法通过氧化除去，只能通过控制废金属原料的化学成分来加以控制，如前面提到的添加直接还原铁和生铁。整个过程中都要保持泡沫渣的存在。泡沫渣经常逸出炉子经炉门流入渣坑。为测定温度和化学成分而进行的取样通过自动氧枪来进行。氧和碳的含量可通过插进钢水的特制探针自动测定，但其他元素的含量则需采取一个"冷样"——一块凝固的小钢样——在电弧发射光谱仪上进行分析。

22　一旦温度和化学成分符合要求，钢水就通过倾斜电炉被倒进一个预热钢包里。如果是普碳钢电弧炉，只要在出钢时发现炉渣，炉子就迅速向除渣侧倾斜回去，最大限度减少下渣到钢包。如果冶炼的是包括不锈钢在内的某些特殊钢种，炉渣会被一起倒进钢包，然后在钢包炉里再处理，回收有价值的合金元素。出钢过程中，某些合金元素会被添加进钢水，更多的石灰被加进钢包上部，开始形成一层新的炉渣。通常会留几吨的钢水和炉渣在电弧炉里，形成"留钢"，帮助预热下次装炉的废金属，加速废金属的熔化。出钢过程中和出钢后，电弧炉要"转回原来的位置"：清除渣门上凝固的炉渣，检查能看见的耐火材料及水冷部件有无泄露，检查电极有无损坏或者给电极添加新的电极段。出钢结束时，需要将出钢口用耐火砂封住。对于一座 90 吨中等功率的电弧炉来说，整个炼钢过程从一炉钢出钢到另一炉出钢中间的时间是 60~70 分钟。

23　电弧炉需要定期将其中的钢和渣完全清空，以便检查耐火炉衬并在必要时对耐火炉衬进行规模较大的修补。因为耐火炉衬通常是用碳酸盐材料煅烧而成，特别容易受到水化作用的影响，因此任何可疑的水冷部件的泄漏都应极其慎重地处理，而不是仅仅关心当前是否有可能发生蒸汽爆炸。耐火炉衬磨损过度会导致漏钢，钢水和炉渣穿透炉衬和炉壳流到周围地区。

电弧炉炼钢的优势

24　电弧炉的使用使得可以百分百采用废金属作原料进行炼钢。与以矿石为原料的一次炼钢相比，这种炼钢流程极大地减少了炼钢所需的能源。电弧炉的另一个优势是灵活：高炉生产不能作很大调整，一次开炉可多年连续运行，而电弧炉可以迅速开始或停止生产，允许钢厂根据需要进行生产调整。虽然电弧炉炼钢通常采用废钢做主要炉料，但如果高炉铁水或直接还原铁可得且在经济上有优势，也可以使用高炉铁水或直接还原铁作电弧炉的炉料。因为电弧炉炼钢需要大量的电力，所以很多公司都利用用电低谷时段电价便宜时安排生产。

25　一座典型的炼钢电弧炉通常为生产棒材或带钢产品的小型钢厂提供钢材。与钢铁联合工厂相比，小钢厂可以离钢产品市场更近，对于运输的要求不高，而钢铁联合工厂大多要求接近港口，可以使用船运。

环境问题

26　虽然现代化电弧炉能够高效地回收利用废钢，电弧炉的运转仍会带来一些负面的环境效应。电弧炉建设中不少基建成本都试图用来减少这些负面效应，包括：

将电弧炉与周围隔开以降低高分贝噪声；

电弧炉废气废尘收集系统

炉渣生产

冷却水需求

用来运送废钢铁、各种材料和产品的重型卡车

发电造成的环境效应。

27　因为电弧炉的负载变化非常大，所以电力系统可能会要采取技术措施才能保证对别的客户的供电质量。电弧炉运转时经常会给电力系统带来闪烁和谐波失真等负面效应。因为这个原因，电站应尽量离电弧炉近一些。

其他种类的电弧炉

28　炼钢(也可以)使用直流电弧炉。直流电弧炉炉顶上只有一个电极，电流经过导电炉底衬或底座的导电销回流。直流电弧炉的优点是生产吨钢消耗的电极较少——因为只使用一个电极，电力谐波及其他类似的问题也较少。但直流电弧炉的(生产)规模受到电极载流量和最大允许电压的限制。导电炉底的维护是扩大直流电弧炉生产规模的瓶颈。

29　钢厂使用钢包炉保持钢水在从电弧炉出钢后处理过程中的温度或改变合金成分。钢包用于保温主要是在离接下来的炼钢流程还有一段时间间隔的情况下。钢包炉包括一个耐火炉顶、一个加热系统以及在条件合适时可将氩气吹进钢水底部用来搅拌的设备。跟废金属冶炼炉不同的是，钢包炉没有倾动机构，也没有废金属装料设备。

30　电弧炉也用来生产碳化钙、铁合金和其他有色合金以及磷。用于这些目的的电弧炉与炼钢用电弧炉在结构上不同，也可能不是一炉炉地生产，而是连续生产。连续生产电弧炉也可能会使用糊状的索德伯格电极(Søderberg electrodes)，防止电极更换造成的生产中断。这样的电弧炉被称为埋弧电炉/潜弧电炉，因为电极尖/电极端部埋入了炉渣/炉料，起弧发生在熔锍和电极之间，穿过炉渣。相比之下，炼钢用电弧炉起弧发生在炉料之外。关键在于电阻，因为是电阻产生需要的热量：炼钢电弧炉中电阻是空气，而在埋弧电弧中，炉渣或炉料形成电阻。两种电弧炉中的液态金属导电太好，无法形成有效的电阻产生热量。

31　业余爱好者建造了各种各样的电弧炉，这些电弧炉常以装在硅石箱子或花盆里的电弧焊接套装为基础。虽然简单、粗糙，这些电弧炉可以熔炼多种多样的材料，生产电

石等。

电弧炉冷却方法

32 无压力水冷却。该工艺也称为喷淋冷却，在大气压力下，使用喷嘴将冷却水喷到电弧炉炉顶、炉壁和排气管道上，实现冷却的目的。每个喷嘴的位置和大小能够保证喷嘴可以根据不同设备不同的热负荷提供需要量的冷却水。喷嘴能沿受热面提供均匀的紊流，是一种有效的冷却方式。该技术由炼钢工作者作为较安全的替代方案研发出来，由联合碳化物公司申请了专利。系统集团的子公司系统喷淋有限公司后来购买了冷区技术专利。

等离子电弧炉

33 等离子使用等离子枪来代替石墨电极。每一枝等离子枪包括一个带喷嘴的套管、一个用来为等离子形成供气(氮气或氩气)的轴向管和一个位于轴向管内部可燃烧的圆柱形石墨电极。

真空电弧重熔

34 真空电弧重熔(VAR)为二次重熔工艺，用于钢锭的真空精炼和制造，以提高其化学和机械均匀性。

35 在一些关键性的军事和商用航天应用领域，材料工程师通常指定要用 VIM-VAR 钢。VIM 指的是 Vacuum Induction Melted(真空感应熔炼)，VAR 指的是 Vacuum Arc Remelted(真空电弧重熔)。VIM-VAR 钢成为了飞机发动机轴承、军用直升机转轴、战斗机襟翼致动器、飞机或直升机传动机构中的齿轮、飞机发动机底座或紧固件、飞机尾钩以及其他一些要求高的应用钢。大多数钢种只冶炼一次，然后被铸成或浇注成固体形状，接着再被广泛锻造或轧制成冶金学上合适的形状。与此对比不同的是，VIM-VAR 钢在真空下经过两次高度纯净的熔炼。电弧炉里冶炼及在氩氧脱碳炉合金化后，准备真空重熔的钢浇铸进钢锭模。凝固的钢锭然后被运往真空感应冶炼炉。通过真空重熔，去除钢中的夹杂物和有害的气体，优化钢的化学成分。VIM 操作在无污染的真空中将这些固体的钢锭还原到液体状态。这种严格控制的冶炼常常需要超过 24 个小时。钢水仍在真空中从 VIM 炉缸流入巨大的电极模。一个典型的电极高约 15 英尺(5 米)，直径不一。电极在真空中凝固。

36 对于 VIM-VAR 钢来说，在下一次真空重熔之前，冷却电极的表面必须打磨，除去表面不规则的地方和杂质。然后打磨过的电极被放进 VAR 炉中。在 VAR 炉的真空密封室里，钢一点点慢慢熔化。真空电弧重熔进一步除去钢中参与的夹杂物，确保钢的超级洁净，同时除去钢中的一些气体，如氧气、氮气和氢气。控制钢水形成液滴及凝固的速度，可以保证整个 VIM-VAR 钢锭化学成分及微观结构的一致。这反过来能让钢更能抗断裂或疲劳。这种精炼过程是满足直升机转轴、军用飞机襟翼致动器或飞机发动机轴承等部件性能特征所必需。

37　为满足某些商业或军事应用需要，钢合金可能只经过一次真空重熔，即 VAR。比如：固体火箭金属壳、战斗机的起落架或扭力杆等用钢，通常只真空重熔一次。

38　真空电弧重熔也用于钛和其他活泼金属或需要较高纯度的金属的生产。

五、专题讨论

外位语与科技翻译

"外位语是用于句外但有内应的成分。具体说，它位于句首或句末，不跟句中一般成分发生结构关系，但跟句中充当结构成分的某一词语指称同一对象。"①例如：

<u>实现四个现代化</u>，这是我们今后相当长时间的中心工作。

在这个例句中，"实现四个现代化"与"这"指称同一对象，形成复指。"实现四个现代化"在结构以外，叫"外位语"（extraposition），"这"在结构以内，称为"本位语"。

使用外位语的目的在于突出某一事物或某一概念，以引起别人的注意，或为了简化复杂的长句，使之结构严谨，条理清楚。前述例句中，使用外位语，就是为了突出"实现四个现代化"这个中心工作。还比如：

<u>正确的东西，好的东西</u>，人们一开始常常不承认它们是香花，反而把它们看作毒草。

这个例句当然也可以表达为"人们一开始常常不承认正确的东西、好的东西是香花，反而把它们看作毒草"，但相比后者，前者结构更简单，意思更清楚。

外位语有外位主语、外位宾语、外位定语等之分。在"实现四个现代化，这是我们今后相当长时间的中心工作"句中，"这"是句子的主语，所以"实现四个现代化"可算作外位主语，而在"正确的东西，好的东西，人们一开始常常不承认它们是香花，反而把它们看作毒草"句中，"它们"分别是"不承认"和"把"的宾语②，所以"正确的东西，好的东西"是

---

① 邢福义，汪国胜. 现代汉语[M]. 武汉：华中师范大学出版社，2006：332.
② 严格说来，这里的"它们"是兼语，既是"不承认""把"的宾语，又是后面句子的主语。事实上，典型的外位宾语还是很多的，如"这孩子，我认识他"。在这个句子中，"这孩子"无疑是外位宾语。

外位宾语。至于下面例句：

现在农村中流行的一种破坏工商业、在分配土地的问题上主张绝对平均主义的思想，它的性质是反动的、落后的、倒退的。

由于"它的"在后面小句中是定语，所以"现在农村中流行的一种破坏工商业、在分配土地的问题上主张绝对平均主义的思想"可看作外位定语。

外位语不是汉语所独有，英语中也有外位语。莎士比亚戏剧 *Hamlet* 中那句著名的 To be or not to be, that is the question, To be or not to be 就是外位语。还比如《圣经》英译本中的这句：

And his sisters, are they not all with us?

很明显，his sisters 也是外位成分。

有些人认为，下面一类英语句子中的不定式短语、动名词或动名词短语，都应看作外位语：

It is a problem how to make both ends meet.
It is wrong to tell a lie.
It is no use talking.
She feels it her duty to help others.
It's dangerous playing with fire. ①

这种观点也有一定的道理。

不过，由于汉语是以话题（topic）为中心来组织句子，英语是围绕主谓结构（subject + predicate）来组织句子，所以总的说来，汉语中带有外位语句子的数量要远远超过英语，例如下面的汉语句子都带有外位语：

枣树，他们简直落尽了叶子。（鲁迅《秋夜》）
那位唐雎老头儿，你和他很熟吧？（郭沫若《虎符》）
他厌恶我，你的父亲……（曹禺《雷雨》）
知识分子，工商业家，宗教家，民主党派，民主人士，必须在反帝反封建的基础

---

① 童养性."外位语"之说当用于英语教学[J]. 外语教学，1983(4)：27.

上将他们团结起来，并加以教育。(毛泽东《中共中央政治局扩大会议决议要点》)

<u>中国一定要坚持改革开放，这是解决中国问题的希望</u>。但是要改革，就一定要有稳定的政治环境。(邓小平《压倒一切的是稳定》)

<u>提出建立社会主义市场经济体制的改革目标</u>，这是我们党在建设中国特色社会主义进程中的一个重大理论和实践创新，解决了世界上其他社会主义国家长期没有解决的一个重大问题。(习近平《切实把思想统一到党的十八届三中全会精神上来》)

谈到外位语，有必要提及英语中的同位语。外位语和同位语有相似之处：它们都和句子中的另一个成分指称相同。两者的区别在于：外位语和相关的代词只是简单地指称同一事物，而同位语则起着补充说明的作用，它与相关的成分包含不同的信息。例如：

Water, a compound, can be decomposed into the gaseous elements hydrogen and oxygen.

在这个句子中，a compound 是 Water 的同位语，补充说明"水"是"一种化合物"。

作为科技翻译工作者，了解外位语可以帮助我们更好地进行翻译，因为汉语一般不使用长句，因此在英译汉时，往往要把原文中较长的句子成分或不易安排的句子成分拆开来，而拆分表达时使用外位结构是个不错的选择。例如：

Water, whether in the Pacific Ocean or in the Atlantic Ocean, consists of hydrogen and oxygen, which is an undeniable scientific fact.

<u>太平洋的水也好，大西洋的水也罢，都是由氢和氧组成的</u>，这是一个不容否认的事实。

The "actions" of the conscious mind, like choosing, whose existence as genuine operations many would deny, still give scientists nightmares, philosophers headaches, and theologians eternal joy.

<u>头脑中有意识的思维活动，比如选择等，尽管许多人不承认这些活动作为实实在在的行为存在着</u>，它们却仍然给科学家带来梦魇，使哲学家大伤脑筋，使神学家从中得到永恒的喜悦。

冯伟年在《"外位语"结构在翻译中的运用》[①]一文归纳总结了英译汉时使用外位结构的八种情况，得出的结论对于英汉科技翻译同样适用。

---

① 冯伟年. "外位语"结构在翻译中的运用[J]. 外语教学(西安外国语学院学报)，1991(4)，52-56.

1. 套用原文的外位语框架。如:

High quality coke is characterized by a definite set of physical and chemical properties that can vary within narrow limits.

高质量焦炭具备<u>一系列较为固定的物理和化学性质</u>,这些性质只在很小的范围内波动。

如果翻译为"高质量焦炭具备一系列较为固定的、只在很小的范围内波动的物理和化学性质",准确性肯定没问题,但译句有太长之嫌,使用外位结构,译文就清楚易懂得多。

CSR measures the potential of the coke to break into smaller size under a high temperature CO/CO_2 environment that exists throughout the lower two-thirds of the blast furnace.

CSR 测量的是焦炭在<u>高温且充满 CO/CO_2</u> 的环境下破碎的可能性,这样的环境出现在高炉下面三分之二的区域。

如果不采用外位语来翻译,译文将非常臃肿。

Basic refers to the magnesia (MgO) refractory lining which wears through contact with hot, basic slags.

"碱性"指的是<u>氧化镁(MgO)耐火炉衬</u>,它经受着高温碱性炉渣的磨损。

Liquid water changes to vapor, which is called evaporation.

<u>液态水变成水蒸气</u>,这就叫蒸发。

2. 把带有某些形容词的句子译成外位语。例如:

The product of this reaction, carbon monoxide, is necessary to reduce the iron ore as seen in the previous iron oxide reactions.

<u>反应生成物一氧化碳为还原铁矿石所必须</u>,这从前面给出的铁氧化物反应式可看出。

3. 把带有某些副词的句子译成外位语。例如:

He made many discoveries in nuclear physics, but characteristically, he did not publish them.

<u>他在原子物理上有许多发现,但并没有发表这些成果</u>,这是他的特点。

这种译法通常发生在句子中的副词大多表达观点或评价的情况下。

4. 把名词和名词短语译成外位语。例如:

Amateurs have constructed a variety of arc furnaces, often based on electric arc welding kits contained by silica blocks or flower pots.

业余爱好者建造了<u>各种各样的电弧炉</u>,这些电弧炉常以装在硅石箱子或花盆里的电弧焊接套装为基础。

The only important ore for aluminum production is bauxite, a mixture of hydrated aluminum oxides ($Al_2O_3 \cdot H_2O$ and $Al_2O_3 \cdot 3H_2O$), which usually contains iron oxide and silica as impurities.

<u>对于铝的生产来说唯一重要的矿石为铝土矿</u>,这是一种水合铝氧化物的混合物($Al_2O_3 \cdot H_2O$ 和 $Al_2O_3 \cdot 3H_2O$),通常含有氧化铁和二氧化硅等杂质。

Wires of different materials, although of the same diameter and length, differ in their conductivity.

<u>不同材料的导线</u>,尽管直径、长度相同,其导电率是不一样的。

可以看出,将名词或名词短语译成外位语,通常发生在名词或名词短语后面存在后置定语或后置状语的情况下,且后置定语或后置状语较长,不使用外位结构不方便表达。

5. 把某些介词短语译成外位语。例如:

When Nucor—now one of the largest steel producers in the US—decided to enter the long products market in 1969, they chose to start up a mini-mill, with an EAF as its steelmaking furnace, soon followed by other manufacturers.

当现今美国最大的钢铁生产商之一的纽柯钢铁公司在1969年决定进入长材市场时,他们选择开班一家小钢厂,<u>用电弧炉做炼钢炉</u>。这一做法很快为别的钢铁生产商效仿。

The melting point is different for different kinds of metals.

<u>不同种类的金属</u>,其熔点各不相同。

6. 用外位语处理 it 为先导词的句型结构。例如:

It is a very important law in physics that energy can be neither created nor destroyed.

能量既不能创造也不能消灭，这是物理学中一条很重要的定律。

7. 把限制性定语从句连同先行词译成外位语。例如：

As in the electron tube, the range of input voltages which may be applied to the transistor is very small.

如同电子管一样，可加给晶体管的输入电压，其范围是很小的。

This property that air has of taking up a large amount of water when cooled is the cause of our cloud and rain.

空气冷却时，会吸进大量的水分，这种特性是云和雨形成的原因。

The electron theory is based on the belief that all matter is electrical in nature.

一切物质在本质上是带电的，这一观点是电子论的基础。

8. 把同位语从句译成外位语。例如：

These efficiencies stem from the fact that in smelting the raw concentrate is used as a valuable fuel to provide the process heat using only oxygen from the air, the process takes place rapidly at a high temperature, and nearly 100% of metal values is recovered, all in a relatively few process steps.

这种高效来自这样的事实：熔炼时，粗精矿本身就被用作有价值的燃料，提供生产需要的热量，只需要使用空气中的氧气即可，熔炼过程在高温下快速发生，金属几乎100%被回收，而所有这一切只需要相对不多的步骤。

The removal of impurities by forming chemical compounds takes advantage of the fact that different metals have different affinities to oxygen, sulfur, or other nonmetallic elements.

通过形成化合物从而去除杂质利用了这样一个事实，那就是不同的金属对于氧、硫和其他非金属元素有不同的亲和力。

既然上述八种类型的英语句子可以翻译成带外位语的汉语句子，汉语带外位语的句子当然也可以用上述八种类型的英语句子来翻译。兹举几例如下：

引起物体运动的任何影响，我们都称之为力。

Any influence that causes a motion of matter we call a force.

一台机器竟能记住这么多事情，而且有求必应，随时待命，她觉得这真是太奇妙了。

It seemed marvelous to her that a machine could remember so much and was always there, ready and waiting to do its work.

20世纪90年代初，中国科学家开始对农作物基因组及其相关领域进行研究，这比国外同行晚了3到5年。

Chinese scientists started crop genome and related research in the early 1990s, three to five years later than their overseas counterparts.

保存太久的食品会变坏，这是因为受到了酵母菌、霉菌和细菌的侵袭。

Food which is kept too long decays because it is attacked by yeasts, moulds and bacteria.

蓝牙是无线能替代电缆的技术，它是基于短波无线电用来连接数字设备的一种技术。

Bluetooth is a wireless cable-replacement technology based on short-range radio used to connect digital devices.

需要指出的是，虽然汉语外位结构在英译汉中非常有用，但由于使用了外位语，多出来的本位语可能会影响译句的简洁和连贯，翻译时需要加以注意。例如：

A movement of air which carries heat from a warmer place to a cooler one is called convection current.

这个句子诚然可以使用外位语翻译成：

带有热量的空气从一个较热的地方向较冷的地方运动，这种运动叫作空气的对流。

但无疑不如下面的译句简洁、紧凑：

带有热量的空气从一个较热的地方向较冷的地方运动叫作空气的对流。

这种简洁、紧凑正是汉语下定义时所需要的。换句话说，如果翻译时希望达到一种简洁、紧凑的效果，那么使用外位语时则须慎重。

◎ 练习八 将下面的句子翻译成汉语。

1. Bearing in mind the contemporary tendencies to implement and develop technologies contributing to protection of natural resources and environment, including those reducing the carbon dioxide emission, an EAF as a scrap-based steel smelting aggregate is a particularly attractive solution.

2. Other considerations include minimization of scrap cave-ins which can break electrodes and ensuring that large heavy pieces of scrap do not lie directly in front of burner ports which would result in blow-back of the flame onto the water cooled panels.

3. The reaction of oxygen with carbon in the bath produces carbon monoxide, which either burns in the furnace if there is sufficient oxygen, or is exhausted through the direct evacuation system where it is burned and conveyed to the pollution control system.

4. Phosphorus and sulfur occur normally in the EAF charge in higher concentrations than are generally permitted in steel and must be removed.

5. Electrical systems in an EAF meltshop usually consist of a primary system which supplies power from the electrical utility; and the secondary electrical system which steps down the voltage from the utility and supplies the power to the EAF.

6. Graphite electrodes are composed of a mixture of finely divided, calcined petroleum coke mixed with about 30% coal tar pitch as a binder, plus proprietary additives unique to each manufacturer.

7. Historically, electrode consumption has been as high as 12—14 pounds per tons of steel, but through continuous improvement in electrode manufacturing and steelmaking operations, this has been reduced to the neighborhood 3.5 to 4.5 pounds per ton.

8. The projection of liquid steel and slag droplets by bursting of CO bubbles has been recognized as the principal mechanism of dust emission in EAF.

9. AC and DC arc furnaces represent one of the most intensive disturbing loads in the sub-transmission or transmission electric power systems; they are characterized by rapid changes in absorbed powers that occur especially in the initial stage of melting, during which the critical condition of a broken arc may become a short circuit or an open circuit.

10. Nowadays, arc furnaces are designed for very large power input ratings and due to the nature of both the electrical arc and the melt down process, these devices can cause large power quality problems on the electrical net, mainly harmonics, inter-harmonics, flicker and voltage imbalances.

第九课 轧制(金属加工)

> 要加强产品结构调整和工艺优化,努力提高自主创新能力,多生产具有市场竞争力的产品。
> ——胡锦涛2005年8月22日考察武钢第二热轧带钢厂时的讲话①

一、科技术语

recrystallization A technique in chemistry used to purify chemicals. By dissolving both impurities and a compound in an appropriate solvent, either the desired compound or impurities can be coaxed out of solution, leaving the other behind.(再结晶)

structural steel A strong steel that is rolled into shapes that are used in construction.(结构钢)

rolling mill A machine or factory in which metal is rolled into sheets or bars.(轧钢厂;轧机)

tandem mill A rolling mill consisting of two or more stands in succession, synchronized so that the metal passes directly from one to another.(连轧机)

continuous casting A technique in which an ingot, billet, tube, or other shape is continuously solidified and withdrawn while it is being poured, so that its length is not determined by mold dimensions.(连铸)

soaking pit A high-temperature, gas-fired, tightly covered, refractory-lined hole or pit into which a hot metal ingot (with liquid interior) is held at a fixed temperature until needed for rolling into sheet or other forms.(均热炉;等温坑)

directionality The property of a microphone or antenna of being more sensitive in one direction than in another.(方向性;定向性)

mill scale Often shortened to just scale, the flaky surface of hot rolled steel, iron oxides

---

① 详参:https://www.gov.cn/jrzg/2005-08/24/content_25665_3.htm。

consisting of iron(Ⅱ,Ⅲ) oxide, hematite, and magnetite.(轧屑;氧化皮)

cross-section A section created by a plane cutting a solid perpendicular to its longest axis.(截面)

surface finish Also known as surface texture, the characteristics of a surface. It has three components: lay, surface roughness, and waviness.(表面光洁度)

ductility The malleability of something that can be drawn into threads or wires or hammered into thin sheets.(延展性;柔软性)

precipitation hardening A process in which alloys are strengthened by the formation, in their lattice, of a fine dispersion of one component when the metal is quenched from a high temperature and aged at an intermediate temperature.(沉淀硬化)

二、英语原文

Rolling (Metalworking)[①]

1 In metalworking, rolling is a metal forming process in which metal stock is passed through one or more pairs of rolls to reduce the thickness and to make the thickness uniform. The concept is similar to the rolling of dough. Rolling is classified according to the temperature of the metal rolled. If the temperature of the metal is above its recrystallization temperature, then the process is known as hot rolling. If the temperature of the metal is below its recrystallization temperature, the process is known as cold rolling. In terms of usage, hot rolling processes more tonnage than any other manufacturing process, and cold rolling processes the most tonnage out of all cold working processes. Roll stands holding pairs of rolls are grouped together into rolling mills that can quickly process metal, typically steel, into products such as structural steel (I-beams, angle stock, channel stock, and so on), bar stock, and rails. Most steel mills have rolling mill divisions that convert the semi-finished casting products into finished products.

2 There are many types of rolling processes, including ring rolling, roll bending, roll forming, profile rolling, and controlled rolling.

History

Iron and Steel

3 The invention of the rolling mill in Europe may be attributed to Leonardo da Vinci in

---

① 本文选自 https://www.mststeel.com/hot-and-cold-rolled-steel-explained/,有删改。

his drawings. The earliest rolling mills in crude form but the same basic principles were found in Middle East and South Asia as early as 600 BCE. Earliest rolling mills were slitting mills, which were introduced from what is now Belgium to England in 1590. These passed flat bars between rolls to form a plate of iron, which was then passed between grooved rolls (slitters) to produce rods of iron. The first experiments at rolling iron for tinplate took place about 1670. In 1697, Major John Hanbury erected a mill at Pontypool to roll "Pontypool plates"—blackplate. Later this began to be rerolled and tinned to make tinplate. The earlier production of plate iron in Europe had been in forges, not rolling mills.

4 The slitting mill was adapted to producing hoops (for barrels) and iron with a half-round or other sections by means that were the subject of two patents of c. 1679.

5 Some of the earliest literature on rolling mills can be traced back to Christopher Polhem in 1761 in Patriotista Testamente, where he mentions rolling mills for both plate and bar iron. He also explains how rolling mills can save on time and labor because a rolling mill can produce 10 to 20 or more bars at the same time.

6 A patent was granted to Thomas Blockley of England in 1759 for the polishing and rolling of metals. Another patent was granted in 1766 to Richard Ford of England for the first tandem mill. A tandem mill is one in which the metal is rolled in successive stands; Ford's tandem mill was for hot rolling of wire rods.

Other Metals

7 Rolling mills for lead seem to have existed by the late 17th century. Copper and brass were also rolled by the late 18th century.

Modern Rolling

8 Modern rolling practice can be attributed to the pioneering efforts of Henry Cort of Funtley Iron Mills, near Fareham, England. In 1783, a patent was issued to Henry Cort for his use of grooved rolls for rolling iron bars. With this new design, mills were able to produce 15 times more output per day than with a hammer. Although Cort was not the first to use grooved rolls, he was the first to combine the use of many of the best features of various ironmaking and shaping processes known at the time. Thus modern writers have called him "father of modern rolling".

9 The first rail rolling mill was established by John Birkenshaw in 1820, where he produced fish bellied wrought iron rails in lengths of 15 to 18 feet. With the advancement of technology in rolling mills, the size of rolling mills grew rapidly along with the size of the

products being rolled. One example of this was at The Great Exhibition in 1851, where a plate 20 feet long, 3½ feet wide, and 7/16 of inch thick, and weighing 1,125 pounds, was exhibited by the Consett Iron Company. Further evolution of the rolling mill came with the introduction of Three-high mills in 1853 used for rolling heavy sections.

Hot and Cold Rolling

Hot Rolling

10   Hot rolling is a metalworking process that occurs above the recrystallization temperature of the material. After the grains deform during processing, they recrystallize, which maintains an equiaxed microstructure and prevents the metal from work hardening. The starting material is usually large pieces of metal, like semi-finished casting products, such as slabs, blooms, and billets. If these products came from a continuous casting operation the products are usually fed directly into the rolling mills at the proper temperature. In smaller operations, the material starts at room temperature and must be heated. This is done in a gas- or oil-fired soaking pit for larger workpieces; for smaller workpieces, induction heating is used. As the material is worked, the temperature must be monitored to make sure it remains above the recrystallization temperature. To maintain a safety factor a finishing temperature is defined above the recrystallization temperature; this is usually 50 to 100℃ (90 to 180℉) above the recrystallization temperature. If the temperature does drop below this temperature the material must be re-heated before more hot rolling.

11   Hot-rolled metals generally have little directionality in their mechanical properties and deformation induced residual stresses. However, in certain instances non-metallic inclusions will impart some directionality and workpieces less than 20 mm (0.79 in) thick often have some directional properties. Also, non-uniform cooling will induce a lot of residual stresses, which usually occurs in shapes that have a non-uniform cross-section, such as I-beams. While the finished product is of good quality, the surface is covered in mill scale, which is an oxide that forms at high temperatures. It is usually removed via pickling or the smooth clean surface (SCS) process, which reveals a smooth surface. Dimensional tolerances are usually 2% to 5% of the overall dimension.

12   Hot-rolled mild steel seems to have a wider tolerance for amount of included carbon than does cold-rolled steel, and is, therefore, more difficult for a blacksmith to use. Also for similar metals, hot-rolled products seem to be less costly than cold-rolled ones.

13   Hot rolling is used mainly to produce sheet metal or simple cross-sections, such as rail

tracks. Other typical uses for hot-rolled metal:

Truck frames;

Automotive clutch plates, wheels and wheel rims;

Pipes and tubes;

Water heaters;

Agricultural equipment;

Strappings;

Stampings;

Compressor shells;

Metal buildings;

Railroad hopper cars and railcar components;

Doors and shelving;

Discs;

Guard rails for streets and highways, etc.

14　Rolling mills are often divided into roughing, intermediate and finishing rolling cages. During shape rolling, an initial billet (round or square) with edge of diameter typically ranging between 100 – 140 mm is continuously deformed to produce a certain finished product with smaller cross section dimension and geometry. Different sequences can be adopted to produce a certain final product starting from a given billet. However, since each rolling mill is significantly expensive (up to 2 million euros), a typical requirement is to contract the number or rolling passes. Different approaches have been achieved including empirical knowledge, employment of numerical models, and Artificial Intelligence techniques. Lambiase et al. validated a finite element model (FE) for predicting the final shape of a rolled bar in round-flat pass. One of the major concern when designing rolling mills is to reduce the number of passes; a possible solution to such requirement is represented by the slit pass also called split pass which divided an incoming bar in two or more subpart thus virtually increasing the cross section reduction ratio per pass as reported by Lambiase. Another solution for reducing the number of passes in the rolling mills is the employment of automated systems for Roll Pass Design as that proposed by Lambiase and Langella. Subsequently, Lambiase further developed an Automated System based on Artificial Intelligence to optimize and automatically design rolling mills.

Cold Rolling

15　Cold rolling occurs with the metal below its recrystallization temperature (usually at room temperature), which increases the strength via strain hardening up to 20%. It also improves the surface finish and holds tighter tolerances. Commonly cold-rolled products include sheets, strips, bars, and rods; these products are usually smaller than the same products that are hot rolled. Because of the smaller size of the workpieces and their greater strength, as compared to hot rolled stock, four-high or cluster mills are used. Cold rolling cannot reduce the thickness of a workpiece as much as hot rolling in a single pass.

16 Cold-rolled sheets and strips come in various conditions: full-hard, half-hard, quarter-hard, and skin-rolled. Full-hard rolling reduces the thickness by 50%, while the others involve less of a reduction. Cold rolled steel is then annealed to induce ductility in the cold rolled steel which is simply known as a Cold Rolled and Close Annealed. Skin-rolling, also known as a skin-pass, involves the least amount of reduction: 0.5%–1%. It is used to produce a smooth surface, a uniform thickness, and reduce the yield point phenomenon (by preventing Lüders bands from forming in later processing). It locks dislocations at the surface and thereby reduces the possibility of formation of Lüders bands. To avoid the formation of Lüders bands it is necessary to create substantial density of unpinned dislocations in ferrite matrix. It is also used to break up the spangles in galvanized steel. Skin-rolled stock is usually used in subsequent cold-working processes where good ductility is required.

17 Other shapes can be cold-rolled if the cross-section is relatively uniform and the transverse dimension is relatively small. Cold rolling shapes requires a series of shaping operations, usually along the lines of sizing, breakdown, roughing, semi-roughing, semi-finishing, and finishing.

18 If processed by a blacksmith, the smoother, more consistent, and lower levels of carbon encapsulated in the steel makes it easier to process, but at the cost of being more expensive.

19 Typical uses for cold-rolled steel include metal furniture, desks, filing cabinets, tables, chairs, motorcycle exhaust pipes, computer cabinets and hardware, home appliances and components, shelving, lighting fixtures, hinges, tubing, steel drums, lawn mowers, electronic cabinetry, water heaters, metal containers, fan blades, frying pans, wall and ceiling mount kits, and a variety of construction-related products.

Processes

Roll Bending

20 Roll bending produces a cylindrical shaped product from plate or steel metals.

Roll Forming

21 Roll forming is a continuous bending operation in which a long strip of metal (typically coiled steel) is passed through consecutive sets of rolls, or stands, each performing only an incremental part of the bend, until the desired cross-section profile is obtained. Roll forming is ideal for producing parts with long lengths or in large quantities.

22 There are 3 main processes: 4 rollers, 3 rollers and 2 rollers, each of which has as different advantages according to the desired specifications of the output plate.

Flat Rolling

23 Flat rolling is the most basic form of rolling with the starting and ending material having a rectangular cross-section. The material is fed in between two rollers, called working rolls, that rotate in opposite directions. The gap between the two rolls is less than the thickness of the starting material, which causes it to deform. The decrease in material thickness causes the material to elongate. The friction at the interface between the material and the rolls causes the material to be pushed through. The amount of deformation possible in a single pass is limited by the friction between the rolls; if the change in thickness is too great the rolls just slip over the material and do not draw it in. The final product is either sheet or plate, with the former being less than 6 mm (0.24 in) thick and the latter greater than; however, heavy plates tend to be formed using a press, which is termed forging, rather than rolling.

24 Often the rolls are heated to assist in the workability of the metal. Lubrication is often used to keep the workpiece from sticking to the rolls. To fine-tune the process, the speed of the rolls and the temperature of the rollers are adjusted.

Ring Rolling

25 Ring rolling is a specialized type of hot rolling that increases the diameter of a ring. The starting material is a thick-walled ring. This workpiece is placed between two rolls, an inner idler roll and a driven roll, which presses the ring from the outside. As the rolling occurs the wall thickness decreases as the diameter increases. The rolls may be shaped to form various cross-sectional shapes. The resulting grain structure is circumferential, which gives better mechanical properties. Diameters can be as large as 8 m (26 ft) and face heights as tall as 2 m (79 in). Common applications include bearings, gears, rockets, turbines, airplanes, pipes, and pressure vessels.

Structural Shape Rolling

26 Also known as shape rolling and profile rolling, it is the rolling and roll forming of

structural shapes by passing them through a rolling mill to bend or deform the workpiece to a desired shape while maintaining a constant cross-section. Structural shapes that can be made with this metal forming process include: I-beams, H-beams, T-beams, U-beams, angle iron, channels, bar stock, and railroad rails. The most commonly rolled material is structural steel; however others include metals, plastic, paper, and glass. Common applications include: railroads, bridges, roller coasters, art, and architectural applications.

27 Structural shape rolling is a cost-effective way of bending this kind of material because the process requires less set-up time and uses pre-made dies that are changed out according to the shape and dimension of the workpiece. This process can roll workpieces into full circles.

Controlled Rolling

28 Controlled rolling is a type of thermomechanical processing which integrates controlled deformation and heat treating. The heat which brings the workpiece above the recrystallization temperature is also used to perform the heat treatments so that any subsequent heat treating is unnecessary. Types of heat treatments include the production of a fine grain structure; controlling the nature, size, and distribution of various transformation products (such as ferrite, austenite, pearlite, bainite, and martensite in steel); inducing precipitation hardening; and, controlling the toughness. In order to achieve this, the entire process must be closely monitored and controlled. Common variables in controlled rolling include the starting material composition and structure, deformation levels, temperatures at various stages, and cool-down conditions. The benefits of controlled rolling include better mechanical properties and energy savings.

三、分析与讲解

碱性氧气炼钢并不是钢铁冶金的终点，冶炼出来的钢水还需要进一步加工。一般来说，转炉冶炼出的钢水经进一步精炼后通过模铸①或连铸，使钢水凝固成型，得到钢锭或铸坯；钢锭或铸坯经轧制（热轧或冷轧）后，得到需要的形状，如型材、线材、板带、钢管

---

① 模铸是一种传统的使钢水凝固成型的方法，把一炉钢水间断地浇注进钢锭模内，待凝固成形冷却后脱模后得到铸锭。模铸每浇一次都要做模具、冷却再脱模，占地面积大，时间比较长生产效率低。而连铸则是将钢水连续地流入中间包，经中间包混合分流后注入结晶器冷却凝固，得到无限长的铸坯，经切割后的铸坯可以直接供轧钢生产使用。连铸简化了生产工艺流程，省去了模铸工艺的脱模、整模、钢锭均热和开坯工序，基建投资可节省40%，占地面积可减少30%，操作费用可节省40%，耐火材料的消耗可减少15%；提高了金属收得率，大幅度减少了钢坯的切头切尾损失，可提高金属收得率约9%；降低了生产过程能耗，可省去钢锭开坯均热炉的燃动力消耗，可减少能耗1/4~1/2；提高了生产过程的机械化、自动化水平。

等。轧钢还可以帮助改善钢的内部质量。本篇课文就是对轧钢的概要性介绍。

第 1 段：(1)汉语中下定义，要么采用"被定义事物+是/指+描述性词语+属概念"的形式，要么采用"描述性词语+属概念+被称为+被定义事物"的形式。翻译第 1 句时，采用任一形式均可。因为句前用的是 metalworking，所以 rolling 不能译成"轧钢"，而必须译成"轧制"。只有轧制的金属是钢坯时，才能译成"轧钢"。后面的 more pairs of rolls 不能译成"几对轧辊"，而应译成"多对轧辊"，因为轧机可能会有多达 12 辊、20 辊甚至 36 辊，不是"几对"可以涵盖的。(2)第 2 句中 concept 字面意思是"概念"，但最好译成"原理"，方便读者理解"轧制"。(3)因为第 4、5 句是对第 3 句的进一步说明，所以译成汉语后，最好将第 3 句后的句号改成汉语的冒号。翻译第 5 句主句时，为突出两种工艺的不同，最好使用表示转折的连词"则"："则被称为冷轧(the process is known as cold rolling)"。(4)倒数第 2 句中的 pairs of 可以译成"一对对"或者"成对成对"。

第 3 段：(1)第 2 句很难按照原句语序翻译，虽然 in crude form 可以译成 rolling mills 的定语，但 but the same basic principles 很难处理。不如将 in crude form but the same basic principles 提出来，单独译成一个小句，放在其他成分译文之后："最早的轧机是在中东和南亚发现的，时间早至公元前 600 年，外形虽粗糙，但基本原理一致。"这样处理，得心应手。(2)第 3 句中 what is now Belgium 不能译成"今天的比利时"，应译为"相当于今天比利时的地区"，这样不会有歧义。(3)第 4 句中 flat bars，一般译为"扁钢"，但根据此处语境，不是钢而是铁，所以应译成"扁材"。(4)最后一句中的 Europe，根据语境，应理解为"欧洲其他地区"，即"欧洲除英国以外的地区"。

第 4 段：第 3 段介绍了纵剪机被用来生产铁条、马口铁，这一段说纵剪机被用来生产 hoops，为了段落间的衔接，在翻译 The slitting mills was adapted to producing hoops 时，最好添加"还"："纵剪机<u>还</u>被改造用来生产箍圈"。

第 5 段：第 1 句中 for both plate and bar iron，明显是 for producing plate and bar iron，所以翻译时要增词"用来<u>生产</u>铁板和铁条的"。

第 6 段：前两句都带有"in+年份"的时间状语，翻译时均<u>应</u>将时间状语放在句前，一方面这是汉语的表达习惯，另外一方面也便于读者抓住时间线索。

第 7 段：Rolling mills for lead 当然可以译为"用于轧制铅的轧机"或"轧制铅的轧机"，但不如在"铅"后面加"金属"两字，这样读来声音饱满一些。科技翻译也要注意音韵美。

第 8 段：(1)第 1 句中的 England，可以译为"英国"，也可以译为"英格兰"，因为 Fairham 在英国的英格兰。考虑到英格兰更具体，译为"英格兰"应该更好。但如果预期读者地理知识较少，则译为"英国"更好。(2)最后一句中的 writers 对应的汉语表达是"作家、作者、撰写者"，但考虑上下文语境，这里最好译成"学者"，也就是从事科技研究或科技史研究的人。

第 9 段：(1)本段围绕 rail rolling 展开叙述，一个关键的问题是：段落中提到的 rolling 到底轧制的是"铁"还是"钢"？John Birkenshaw 于 1820 年通过轧制生产的 rail 无疑是铁轨（wrought iron），那么后面提到的 rolling mill 呢？是"轧钢厂"还是"轧铁厂"？查阅相关网络资源，应是"轧铁厂"。(2)The Great Exhibition 是专有名词，不可直译。专有名词翻译必须首先查找有无约定俗成的译法，如果有，则必须按照约定俗成的译法进行翻译。查相关资料，The Great Exhibition 的全称是 The Great Exhibition of the Works of Industry of all Nations，一般译作"万国工业博览会"。

第 10 段：(1)因为文章第 1 段就已经给 hot rolling 下过定义了，所以第 1 句就不要用下定义的句式来翻译，可以译成"作为一种金属加工工艺，热轧时，温度高过材料的再结晶温度"。翻译时，表达方式的选择需要考虑上下文情况。(2)in small operations 如果译成词组，表达不会很顺畅，不如译成条件从句"如轧制规模较小"。(3)翻译倒数第 3 句 the temperature must be monitored 时，最好在"温度"后加上"的变化"三字。(4)最后一句条件从句虽然有强调的意思，但翻译时无须强调，平实陈述即可。

第 11 段：第 2 句中的 directional properties，应译为"性能上的方向性"。

第 14 段：(1)第 2 句中 cross section dimension and geometry 是并列关系，但无疑 cross section dimension 也是 geometry（即 geometric parameters）的一种，所以译成汉语时，须将 cross section dimension 译成 geometry 的一个方面："截面尺寸等几何参数"。(2)倒数第 3 句虽然是一句话，但事实上是两句话，只是这两句话关系紧密，所以用分号连接起来，其实将分号改成句号也是可以的。译成汉语时，最好译成两句话，这种拆译技巧是科技翻译中常用的。需要注意的是，as reported by Lambiase 可以理解为修饰 increasing the cross section reduction ratio per pass，也可以理解为修饰 a possible solution to such requirement。结合前文，我们倾向于认为是后者："解决这个问题的一个办法是兰姆拜斯提出的……"。

第 15 段：(1)同热轧，第 1 句也不要采用汉语的定义句式来翻译，因为文章在第 1 段就已经给"冷轧"下了定义。(2)第 2 句中的 hold tighter tolerances 无法直译，可意译为"实现更高精度的轧制"，是对 tolerances（公差）的正说反译。

第 16 段：(1)为了前后衔接，第 3 句中的 then 需要译成"冷轧之后"。(2)倒数第 2 句中的 spangles in galvanized steel，虽然介词用的是 in，但 spangles 无疑在 galvanized steel 表面上，因为平整轧制用来使金属表面光洁。另外，"锌花"（spangles）如果在镀锌钢内部看不见，就无所谓"花"了——"花"必然是有形的。《英汉钢铁冶金技术详解词典》也明确给出 spangles 的释义为"锌花（镀锌薄钢板表面上的）"。

第 17 段：(1)Other shapes 指的是 Other shapes of stock。(2)千万不能误将 sizing, breakdown, roughing, semi-roughing, semi-finishing, finishing 当成普通词语来译，必须根据这些词语的专业释义来译。

第18段：注意主语的结构 the smoother, more consistent, lower levels of carbon encapsulated in the steel，而不是 the smoother, the more consistent, the lower levels of carbon encapsulated in the steel。前者表明 smoother, more consistent, lower 修饰的都是 levels of carbon encapsulated in the steel，而在后一种结构中，the smoother, the more consistent, the lower 修饰的就不是一样东西了，最大可能 the smoother 描写的是"钢材表面"，the more consistent 描写的是"钢材的各部分"。

第20段：原文的 from plate or steel metals 本身表达不是很准确，合理的说法应该是 from plate of steel or other metals，翻译时应按照逻辑进行表达："包括钢在内的金属板材"。

科技翻译时，也会遇到原文不符合逻辑的地方。这时有两种处理方式：一是按照原文翻译，但须标注原文就是如此，避免让译文读者以为译者翻译中犯了错误；二是更正原文的表达错误，按照合乎逻辑的方式进行翻译。

第21段：第1句中 each performing only an incremental part of the bend 可直译为"每组轧辊只完成弯曲的一个增量部分"，也可意译为"每组轧辊只完成某一方向的弯曲"。采用何种译法，取决于读者的专业理解能力。

第24段：翻译第2句时，为了衔接前后，不妨加个"也"字："也常使用润滑，防止工件粘在轧辊之上"。

第25段：(1)翻译倒数第2句时，需要补充 Diameters 的主体"环轧产品的直径"。(2)最后一句翻译，也需补充"环轧"，最后还需要补充"的生产"三个字："环轧常见应用包括轴承、齿轮、火箭、涡轮机、飞机、管道和压力容器的生产"。

第26段：(1)翻译第1句时，须用括号带出 shape rolling, profile rolling 两个术语，因为这两个术语对应的汉语词语都是"型材轧制"。如果只单纯译成"型材轧制"，读者就不会知道在英语中有两个表达都有"型材轧制"的意思。

第27段：因为第2、3句主语都是 This process，所以第1句的 Structural shape rolling 最好译成"结构板型轧制工艺"，后面的 this kind of material 最好译成"这类材料"，比"这种材料"涵盖范围广，因为前面提到除结构钢外，其他金属、塑料、纸张和玻璃都可以进行结构板型轧制，所以材料的范围比较广。

第28段：翻译第2句时，为前后衔接，最好在句前添加"控制轧制时"五个字。同样，在翻译后面的 this entire process must be closely monitored and controlled 时最好也添加"控制轧制"四个字："整个控制轧制过程都必须严密监控"。

四、参考译文

轧制(金属加工)

1　金属加工中,金属原料通过一对或多对轧辊使其厚度变薄而均匀的金属成型工艺被称为轧制。轧制的原理同擀面团类似。轧制根据轧制金属的温度进行分类:如果金属的温度高于其再结晶温度,这种轧制被称为热轧;如果金属的温度低于其再结晶温度,则被称为冷轧。就使用范围而言,热轧处理的金属比任何其他制造工艺处理的都多,而在所有冷加工工艺中,冷轧处理的技术为最多。夹持一对对轧辊的轧机机架组合在一起成为轧机。轧机能迅速处理金属,尤其是钢,将其加工成结构钢(工字钢、角钢、槽钢等)、棒材和钢轨。大部分钢厂都有轧钢部,将半成品的铸件产品加工成制成品。

2　轧制工艺有多种,包括环轧、轧弯、轧制成形、型材轧制和控制轧制。

历史

铁和钢

3　轧机在欧洲的发明也许应归功于列奥纳多·达·芬奇的绘画。最早的轧机是在中东和南亚发现的,时间最早可追溯至公元前600年,外形虽粗糙,但基本原理一致。最早的轧机为纵剪机,1590年从相当于今天比利时的地区传到英国。这种轧机将扁材输送到轧辊间,得到铁板,铁板再通过带槽轧辊/槽辊(纵剪机),生产出铁条。最早将铁轧制用于生产马口铁的实验发生在1670年。1697年,约翰·汉伯里少校在庞蒂浦建立一家轧制厂,生产"庞蒂浦板",即黑铁板。后来,这种板被再轧制,镀上锡,称为镀锡板(马口铁)。欧洲(其他地区)较早生产铁板不是在轧制厂,而是在锻造厂。

4　纵剪机还被改造用来生产箍圈(用于生产桶装物品)和半圆或其他断面形状的铁板,这得益于大约1679年的两项专利。

5　一些最早的关于轧机的文献可追溯到克里斯托弗·波勒姆,1761他在《爱国者圣约》中提及用于生产铁板和铁条的轧机。他还解释了轧机如何能够节省时间和劳力,因为一台轧机可同时生产10到20根铁条。

6　1759年,一项金属抛光和轧制的专利被授予了英国的托马斯·布洛克。1766年,另一项也是世界首个连轧机专利被授予了英国的理查德·福特。连轧机可在连续机架上轧制金属。福特的连轧机用于热轧线材。

其他金属

7 用于轧制铅金属的轧机似乎在17世纪晚期已经存在。18世纪晚期，铜和黄铜也已被轧制。

现代轧制简述

8 现代轧制实践可归功于亨利·科特在靠近英格兰英国法汉姆建立的富特利铁厂。1783年，亨利·科特因使用带槽轧辊轧制生产铁条而获批一项专利。使用这种新的设计，轧机较铁锤能够增加日产量15倍。虽然科特不是第一个使用带槽轧辊，但他是第一个综合使用当时已知各种炼铁和成型工艺中很多最好技术的人。因此，当代学者将其称为"现代轧制之父"。

9 第一座铁轨轧制厂于1820年由约翰·伯肯肖建成，工厂生产15~18英尺（4.575~5.49米）长的鱼肚形熟铁轨道。随着轧制技术的发展，轧制产品的尺寸在增加，轧制厂的规模也迅速变大。一个例证便是1851年举行的万国工业博览会。博览会上，康塞特制铁公司展示了一块20英尺（约6.1米）长、3.5英尺（约1.1米）宽、7/16英寸（约1.1厘米）厚的铁板，铁板重达1125磅（约510.3千克）。1853年，随着三辊轧机开始用于轧制大型型材，轧制厂迎来了新的发展。

热轧和冷轧

热轧

10 作为一种金属加工工艺，热轧时，温度高过材料的再结晶温度。热轧中，（金属）颗粒变形，再结晶，保持一种等轴的微观结构，防止金属加工硬化。热轧的初始原料通常是大件的金属，如像板坯、大方坯、小方坯等铸造半成品。如果这些原料来自于连铸生产线，通常原料会在适宜温度时就直接被送进轧制厂。如轧制规模较小，原料处于室温状态，则原料需先加热。加热较大的工件，使用气体或石油做燃料的均热炉；加热较小的工件，则使用感应加热。材料加工过程中，需监控温度（的变化），确保温度始终高于材料的再结晶温度。为了确保轧制效果，（人们）确定了终轧温度，常高于再结晶温度50℃至100℃（90℉ to 180℉）。如果温度低于此温度，则需要重新加热材料，然后再热轧。

11 热轧金属的机械性能通常无方向性，存在形变导致的残余应力。然而，在某些情况下，非金属夹杂物会（给金属）带来某种程度的方向性，厚度小于20mm的工件通常都有某种（性能上的）方向性。此外，冷却不均匀也会产生很多残余应力，这种现象常出现在截面不均匀形状的金属上，如工字钢。如成品质量不错，成品表面会被轧屑覆盖，这是高温下形成的氧化物。通常通过酸洗或光滑清洁表面（SCS）过程去除，去除后呈现光滑的表面。尺寸公差通常为整体尺寸的2%—5%。

12　热轧低碳钢较冷轧钢似乎在含碳量上有更大的容忍度，因此难于被铁匠使用。对于类似金属来说，热轧产品似乎比冷轧产品便宜。

13　热轧主要用于生产薄板或简单的截面，如铁轨。热轧金属的其他典型用途还有：车架，汽车离合器片、车轮和轮辋，管道，热水器，农业设备，带卷，冲压件，压缩机外壳，金属建筑，铁路漏斗车和火车车厢组件，门和棚架，圆盘，街道和高速公路的护栏路的栏杆，等等。

14　轧机通常可分为粗轧、中轧和精轧机组。型材轧制时，边缘直径常在100至140mm的初始坯料(圆形或方形)持续不断地变形，生产出截面尺寸等几何参数较小的某种成品。从给定的坯料到生产出某种最终产品，可以有不同的生产次序。但是，因为轧机很贵(一台高达200万欧元)，(设计生产次序时)典型的要求是减少轧机的数量或轧制道次。(对此)，人们提出了不同的方法，包括使用经验知识、采用数值模型和运用人工智能技术。兰姆拜斯等(提出并)验证了一种有限元模型，用来预测一根轧制的棒材在圆-扁平道次(轧制后)的最终形状。设计轧机时的一个主要考虑是要减少轧制道次。解决这个问题的一个办法是兰姆拜斯提出的使用切分道次，也就是切分轧制，将进来的棒材分成两个或多个部分(进行轧制)，这样做实际上增加了每道次横截面的压下率。另外一个减少轧机道次的办法是使用兰姆拜斯和兰杰拉提出的轧制道次自动设计系统。兰姆拜斯后来还基于人工智能进一步研发了一个自动系统以优化和自动设计轧机。

冷轧

15　冷轧时，金属温度低于其再结晶温度(通常在室温进行)。冷轧通过应变硬化提高金属强度达20个百分点，同时改善金属表面光洁度，实现更高精度的轧制。常见的冷轧产品包括板材、带材、棒材、线材。这些产品冷轧时常较其热轧时尺寸要小。因为与热轧原料相比，冷轧工件尺寸较小，强度更大，因此(冷轧时)使用四辊轧机或多辊轧机。冷轧一个轧制道次降低的工件厚度比热轧要小。

16　板材和带材冷轧时情形不一：全硬、半硬、四分之一硬和平整轧制。全硬冷轧减少工件厚度50%，其他类型的轧制减少的工件厚度要小一些。冷轧之后，冷轧钢需要退火处理，以增加其延展性。这种钢被简单地称为冷轧密闭退火钢(冷轧连续退火钢)。平整轧制也叫光整冷轧，压下量最小，在0.5%到1%之间，用来使表面光洁、厚度均匀，减少屈服点现象(防止在后期的加工过程中形成吕德斯带)。平整轧制将位错锁定在金属表面，因此能够降低吕德斯带形成的可能性。为防止吕德斯带的形成，还需要在铁素体基体上产生密度很大的未钉扎位错，用来打散镀锌钢(表面的)锌花。光整轧制料常用在后续的需要高延展性的冷轧处理中。

17　其他形状的坯料只要截面相对均匀、横向尺寸相对较小，也可以冷轧。冷轧成型涉及一系列的成型处理，通常是按这样的顺序：定径、开坯、粗轧、半粗轧、半精轧和精轧。

18 如果铁匠要加工钢材，则钢中含碳量越低、越均匀、变化越小就越容易，但代价是成本更高。

19 冷轧钢的典型用途包括金属家具、书桌、文件柜、餐桌、椅子、摩托车排气管、计算机机柜和硬件、家用电器和零部件、搁板、照明用固定装置、铰链、管道、钢桶、割草机、电子柜、热水器、金属容器、风扇叶片、煎锅、墙壁和天花板安装套件以及很多与建筑有关的产品。

轧制工艺

辊弯

20 辊弯弯曲包括钢在内的金属板材，以生产一种圆柱形的产品。

辊弯成型

21 辊弯成型为连续弯曲操作。辊弯成型时，一块长的金属带材（通常为带钢）通过连续几组轧辊或者说机架，每组轧辊只完成弯曲的一个增量部分（每组轧辊只完成某一方向的弯曲），直至获得期望的截面轮廓。辊弯成型非常适合生产长尺寸的零件或大批量的零件。

22 辊弯成型有三种工艺：四辊、三辊和两辊，根据产出板材的期望规格，每种工艺有着不同的优点。

平辊轧制

23 平辊轧制是最基本的一种轧制。平辊轧制时，初始材料和制成材料都有着长方形的截面。初始材料被喂入两根被称为工作辊的轧辊间，两根轧辊朝相反方向旋转。两根轧辊之间的空隙比初始材料的厚度要小，使得初始材料变形。材料厚度下降使得材料拉长，材料和轧辊间的摩擦力推动材料穿过轧辊。一个道次轧制可能的变形的量受到轧辊间摩擦力的限制。如果期望厚度的变化太大，轧辊可能会直接滑过材料而无法将材料拉进去。最终产品要么是薄板，要么是中厚板。前者厚度小于6mm，后者厚度大于6mm。厚板常用压锻机成型，称为锻造，而非轧制。

24 通常会将轧辊加热，提高金属的可加工性。也常使用润滑，防止工件粘在轧辊之上。通过调整轧制的速度和轧辊的温度，可以对轧制进行微调。

环轧

25 环轧是一种特殊类型的热轧，目的是增加金属圆环的直径。初始材料为一带厚壁的圆环。工件被放在两根轧辊之间，一根内惰辊/空转辊和一根从动轧辊，从动轧辊从外压制圆环。随着轧制的进行，圆环壁厚降低，直径增加。为轧制出不同截面形状的产品，轧辊的形状也可以改变。产生的晶粒组织呈周向，力学性能更好。（环轧产品）直径可高达8m(26ft)，面高可达2m(79in)。环轧常见应用包括轴承、齿轮、火箭、涡轮机、飞机、管道和压力容器的生产。

结构板形轧制

26　这种轧制也称为型材轧制（shape rolling 或 profile rolling）。轧制时，结构型材通过轧机轧制、辊轧成形，弯曲或变形至期望形状，同时截面保持不变。可使用这种金属成形工艺生产的结构型材包括：工字钢、H型梁、丁字梁、U型钢、角钢、槽钢、条材和铁路钢轨。最常用的轧制材料为结构钢，然而，其他材料如金属、塑料、纸张和玻璃也可以轧制。常见的用途包括铁路、桥梁、过山车、艺术和建筑。

27　结构板形轧制工艺用来轧制这类材料经济划算。这种工艺根据工件形状和尺寸改变磨具，使用预制磨具，因此需要的启动时间较短。这种工艺可将工件轧制成完整的圆的形状。

控制轧制

28　控制轧制是一种将控制变形和热处理结合在一起的热机械加工。控制轧制时，将工件温度提升到再结晶温度之上的热量同时也用来热处理，所以后续就没必要再进行任何热处理了。热处理的类型包括产生细晶结构；控制不同的转变产物的性质、大小和分布（如钢中的铁素体、奥氏体、珠光体、贝氏体、马氏体）；诱导产生析出硬化以及控制韧性。为了实现这一点，必须严密监控整个控制轧制过程。控制轧制中常见的变量包括初始材料的成份和结构、变形程度、各阶段的温度、冷却环境。控制轧制的好处包括更好的力学性能、节省能源。

五、专题讨论

科技英语中状语从句的翻译

英语中，用来修饰动词、副词和形容词的从句叫作状语从句。状语从句根据状语的意思可分为时间状语从句、地点状语从句、条件状语从句、原因状语从句、结果状语从句、比较状语从句、目的状语从句和让步状语从句等。

科技英语揭示事物和概念间的联系，自然地，状语从句在科技英语中使用得十分频繁，起到说明事件发生的时间、地点、条件、方式、因果等重要作用。

科技英语中状语从句翻译的难点，在于如何让原文状语从句传达的信息在汉语中以合适的形式、成分和位置出现，实现相同的信息功能、句法功能以及交际功能。下面通过一些翻译实例，探讨一下科技英语状语从句的汉译问题：

<u>Although Cort was not the first to use grooved rolls</u>, he was the first to combine the use of many of the best features of various ironmaking and shaping processes known at the time.

<u>虽然科特不是第一个使用带槽轧辊的人</u>，但他是第一个综合使用当时已知各种炼铁和成型工艺技术中最好的人。

<u>Once a blast furnace is started</u> it will continuously run for four to ten years with only short stops to perform planned maintenance.

<u>一旦高炉启动</u>，它将连续运行四到十年，只会短暂停歇以执行有计划的维护。

科技文献不乏叙事性内容，比如介绍技术背景或者历史发展脉络的文字。这种语境下的状语从句，一般按正常语序、内容翻译即可，比如上例中的让步状语从句和条件状语从句。还比如 If these products came from a continuous casting operation the products are usually fed directly into the rolling mills at the proper temperature. 可以直译为："如果这些原料来自连铸生产线，通常原料会在适宜温度时就直接被送进轧制厂。"

This ore is crushed and ground into a powder <u>so the waste material called gangue can be removed</u>.

将这种矿石碾碎并磨成粉末，<u>通过这种方式可以将被称为脉石的废物去除</u>。

<u>As the oven is heated</u> the coal is cooked so most of the volatile matter such as oil and tar are removed.

<u>给焦炉加热过程中</u>，煤炭受热，<u>其间大部分挥发性物质被除去</u>，如油和焦油。

科技文献中，时间状语从句和结果状语从句尤其多见。由于汉语讲究"意合"，因此翻译时间状语从句时，较少使用"当/在……时"，而采用较隐蔽的表达方式，暗含时间意义。例如，"给焦炉加热过程中"，不但完整地表达了时间意义，而且读起来简洁自然。为了使译文连贯，有时也会适当地增加词语，如上述例句中的"通过这种方式"和"其间"，都起到了不错的衔接作用。

While the finished product is of good quality, the surface is covered in mill scale, which is an oxide that forms at high temperatures.

成品质量虽然很好，但其表面覆盖着轧屑。这是在高温下形成的氧化物。

Structural shape rolling is a cost-effective way of bending this kind of material because the process requires less set-up time and uses pre-made dies that are changed out according to the shape and dimension of the workpiece.

结构板形轧制工艺用来轧制这类材料经济划算。这种工艺根据工件形状和尺寸，更换预制磨具，因此需要的启动时间较短。

表示转折或因果关系的状语从句在科技英语中不在少数，需要根据具体语境，确定具体的翻译策略。比如上述两例，第一例是 while 引导的让步状语，按正常顺序翻译即可；而第二例中 because 引导的原因状语从句，翻译时就需要稍作调整。由于 because 后面的成分过于复杂，用一个"因为"统摄会非常臃肿，不符合汉语表达习惯，因此最好采取分译法，将状语从句部分译成一个单独的句子，使之与原来的主句部分具有了相同的句法地位。

总体而言，科技英语中的状语从句大多好辨识、易理解、基本不容易译错，需要注意的有两种情况：

一是时间状语从句表示条件关系，需要译成条件句。例如：

Turn off the switch when anything goes wrong with the machine.
如果机器发生故障，就把电闸关上。
A body at rest will not move till a force is exerted on it.
若无外力作用，静止的物体就不会移动。

二是将时间、原因状语从句译成并列的分句。例如：

The earth turns round its axis as it travels around the sun.
地球一面绕太阳运转，一面绕地轴自转。
Where there is sound, there must be sound waves.
哪里有声音，哪里就有声波。
It might have rained last night, for the ground is wet.
地上是湿的，昨晚可能下雨了。

之所以将时间、原因状语从句译成并列的分句，同样是因为前面所说的汉语是"意合"的语言，不喜欢使用过多的关系词，习惯让读者根据句子意思去揣摩、领会句子各部分之间的关系。

◎ 练习九　请将下面的句子翻译成汉语。

1. It is important to improve the electrical machines efficiency for energy conservation. Accordingly, many researchers are committed to the development and production of non-oriented silicon steel for highefficiency motor with high magnetic induction and low core loss.

2. In order to control the grain size before cold rolling and produce high-performance non-oriented silicon steels, one-stage cold rolling process with hot band normalization and two-stage cold rolling process with intermediate annealing are studied comparatively.

3. Although two-stage rolling method will increase the production cost a little, it is conductive to produce high-performance non-oriented electrical steels and has certain potential in some specific application fields.

4. The rapid growth of grains was due to the increase of grain boundary mobility at relatively higher temperature. In addition, the heterogeneous microstructures were attributed to the inheritance of initial microstructure.

5. With the decrease of hot rolling temperature from 1,000℃ to 800℃, the uniform equiaxed prior austenite grains microstructures changed to heterogeneous microstructures of laminated and ultrafine prior austenite grains arranged alternatively.

6. High-strength ultra-heavy plate steels are widely used in architectural structures, bridges, and offshore platforms. In recent years, low-C, medium-Mn transformation-induced plasticity (TRIP) steels have attracted wide attention as a good candidate for advanced heavy steel plates due to their excellent combination of high strength and high ductility.

7. At the moment, research on medium manganese steel is mostly focused on chemical composition and heat treatment parameters, with few investigations on the initial microstructure at various hot rolling temperatures.

8. The width of lamellar retained austenite increased with the decrease in rolling temperature. It showed that Mn diffused more fully in the intercritical annealing process with the decrease in rolling temperature. This was because the Mn element diffused along the dislocation and grain boundary channels, while the R800-IA sample had a high dislocation density due to the low rolling temperature.

9. The energy of atoms at 800℃ was lower than that at 900℃ and 1,000℃, and the recovery and recrystallization at 800℃ were suppressed. Therefore, part of the prior austenite grains maintained elongated morphology after the rolling process and became martensite in subsequent directly water-quenching.

10. In this new corrugated rolling process, the upper corrugated roll contacts metal 1 whose deformation resistance is great, and the lower flat roll contacts metal 2 whose deformation resistance is small. Through this new rolling process, two or more different metals can be processed into corrugated composite plates whose upper surface and bonding surfaces are corrugated.

11. Rolling mill is the main equipment in this process, whose automatic control is the part with the highest automation requirements in the field of steel rolling. The determination of rolling force is of great significance in rolling production, which can provide basis for the setting of roll gap and guide the design or selection of bearing capacity and strength check of equipment.

12. In addition to the dynamic rolling forces, the specific rolling pressure distributions and horizontal stress distributions along the projected contact arc can be easily obtained through the proposed model, which is one of the advantages of building a mathematical model with the slab method.

第十课 有色金属：生产和历史

> 把64种有色金属当64个堡垒，要一个一个地攻下来。
>
> ——王鹤寿①

一、科技术语

amalgam Any of various alloys of mercury with other metals, especially an alloy of mercury and silver used in dental fillings or an alloy of mercury and tin used in silvering mirrors. (汞合金,尤其是银汞或锡汞合金)

anode The positive electrode in the electrolytic cell or through which current enters an electric device.(正极)

arsenic A grayish white element having a metallic luster, vaporizing when heated, and forming poisonous compounds. Symbol：**As**；atomic weight：74.92；atomic number：33.(砷)

assay Qualitative or quantitative analysis of a metal or ore to determine its components.(化验)

bath smelting A smelting method in which physicochemical change in the metallurgical process mainly takes place in the bath (melt) of furnaces in metallurgical industry.(熔池熔炼)

bauxite A sedimentary rock with a relatively high aluminum content.(铝土矿，铝矾土)

(the) Bayer process The principal industrial means of refining bauxite to produce alumina (aluminum oxide) and was developed by Carl Josef Bayer.(拜耳法)

bismuth A chemical element with the symbol **Bi** and atomic number 83. Bismuth is the most naturally diamagnetic element and has one of the lowest values of thermal conductivity among metals.(铋)

cadmium A chemical element, atomic number 48, atomic weight 112.411, symbol **Cd**.(镉)

---

① 这一表述出自时任冶金工业部部长王鹤寿在1958年3月15日向毛泽东主席和党中央提交的报告。

calcine To heat (a substance) to a high temperature but below the melting or fusing point, causing loss of moisture, reduction or oxidation, and the decomposition of carbonates and other compounds.(煅烧)

carbonate Chemical compound containing the carbonate radical or ion, CO_3^{-2}.(碳酸盐)

castability The ability of materials to set in a mould when mixed with water and a bonding agent.(可铸性)

cathode The negative electrode in the electrolytic cell or through which current enters an electric device.(负极)

chromium A metallic chemical element atomic number 24, atomic weight 51.996, symbol Cr.(铬)

cinnabar A heavy reddish mercuric sulfide, HgS, that is the principal ore of mercury.(硫化汞;朱砂)

cobalt A chemical element, atomic number 27, atomic weight 58.933, symbol Co.(钴)

cryolite Also called Greenland spar, an uncommon, white, vitreous natural fluoride of aluminum and sodium, Na_3AlF_6, nearly invisible in water in powdered form and used chiefly in the electrolytic recovery of aluminum.(冰晶石)

cupellation The process of recovering precious metals from lead by melting the alloy in a cupel and oxidizing the lead by means of an air blast; the manufacture of lead oxide by melting and oxidizing lead.(灰吹法;烤钵冶金法)

cyclone smelting A type of metallurgical reaction in the swirl of a vortex room.(漩涡熔炼)

electrolytic cell A device containing two electrodes immersed in a solution of electrolytes, used to bring about a chemical reaction. Electrolytic cells require an outside source of electricity to initiate the movement of ions between the two electrodes where the chemical change takes place.(电解池)

electrorefining Purifying metals by electrolysis using an impure metal as anode from which the pure metal is dissolved and subsequently deposited at the cathode. Also known as **electrolytic refining**.(电解精炼)

electrorefining Purifying metals by electrolysis using an impure metal as anode from which the pure metal is dissolved and subsequently deposited at the cathode. Also known as electrolytic refining.(电解精炼)

flotation A process to concentrate the valuable ore in low-grade ores. The ore is ground to a powder, mixed with water containing surface-active chemicals, and vigorously aerated. The bubbles formed trap the required ore fragments and carry them to the surface froth, which is then skimmed off. Also called **froth flotation**.(浮选;泡沫浮选)

fluidized bed A physical phenomenon occurring when a quantity of a solid particulate substance (usually present in a holding vessel) is placed under appropriate conditions to cause a solid/fluid mixture to behave as a fluid. This is usually achieved by the introduction of pressurized fluid through the particulate medium. This results in the medium then having many properties and characteristics of normal fluids, such as the ability to free-flow under gravity, or to be pumped using fluid type technologies.(流化床)

fractional distillation The process of separating the constituents of a liquid mixture by heating it and condensing separately the components according to their different boiling points.(分馏)

galena A gray mineral, essentially PbS, the principal ore of lead.(方铅矿)

germanium A chemical element, atomic number 32, atomic weight 72.59, symbol **Ge**.(锗)

gravity separation Separation of immiscible phases (gas-solid, liquid-solid, liquid-liquid, solid-solid) by allowing the denser phase to settle out under the influence of gravity; used in ore dressing and various industrial chemical processes.(重力分选)

(the) **Hall-Héroult cell** The device used for the Hall-Héroult process to obtain aluminum.(霍尔—埃鲁特电解槽)

(the) **Hall-Héroult process** The major industrial process for smelting aluminum. It involves dissolving aluminum oxide (alumina) (obtained most often from bauxite, aluminum's chief ore, through the Bayer process) in molten cryolite, and electrolysing the molten salt bath, typically in a purpose-built cell. The Hall-Héroult process applied at industrial scale happens at 940–980℃ and produces 99.5%–99.8% pure aluminum. Recycled aluminum requires no electrolysis, thus it does not end up in this process. This process contributes to climate change through the emission of carbon dioxide in the electrolytic reaction.(霍尔-埃鲁特法)

hydrophilic Having an affinity for water; readily absorbing or dissolving in water.(亲水的; 吸水的)

in situ In the original position.(原位)

indium A chemical element, atomic number 49, atomic weight 114.82, symbol **In**.(铟)

induction heater An electromagnetic device used for induction heating. An induction heater consists of two basic parts: a primary coil, which generates an alternating magnetic field, and current feeders, which connect the primary coil to the power supply. An electrically conductive object placed in an alternating magnetic field heats up as a result of the thermal action of eddy currents induced in the parts of the object that are directly enveloped by the primary coil.(感应加热器)

(the) **Iodide process** A refining process in which a metal, such as titanium or zirconium, is combined with iodine vapor and then the iodide volatilized and decomposed at high temperatures to yield a pure solid metal.(碘化物法)

(the) **Kennecott-Qutotec flash converting process** A process for flash converting characterized by the continuous and closed operation as well as the decoupling of the matte smelting and converting steps.(肯尼科特—奥托昆普闪速吹炼工艺).

(the) **Kroll process** A pyrometallurgical industrial process used to produce metallic titanium from titanium tetrachloride. In the Kroll process, the $TiCl_4$ is reduced by liquid magnesium to give titanium metal: $TiCl_4 + 2Mg \rightarrow Ti + 2MgCl_2$.(克罗尔法)

lithium A chemical element with the symbol Li and atomic number 3. It is a soft, silvery-white alkali metal. Under standard conditions, it is the lightest metal and the lightest solid element.(锂)

lixiviant A liquid medium used in hydrometallurgy to selectively extract the desired metal from the ore or mineral.(浸出剂)

magnetic separation The process of separating components of mixtures by using magnets to attract magnetic materials. The process that is used for magnetic separation detaches non-magnetic material with those that are magnetic. During magnetic separation, magnets are situated inside two separator drums which bear liquids. Due to the magnets, magnetic particles are being drifted by the movement of the drums. This can create a magnetic concentrate (e.g. an ore concentrate).(磁选)

malachite A mineral, the green basic carbonate of copper occurring in crystals of the monoclinic system or (more usually) in masses.(孔雀石)

manganese A metallic element, atomic weight 54.938, atomic number 25, symbol Mn.(锰)

(the) **Mond process** A process for extracting and purifying nickel whereby nickel carbonyl is first formed by reaction of the reduced metal with carbon monoxide, and then the nickel carbonyl is decomposed thermally, resulting in deposition of nickel.(蒙德法)

niobium Also known as columbium, a chemical element with the symbol Nb (formerly Cb) and atomic number 41. It is a light grey, crystalline, and ductile transition metal.(铌)

nonferrous metal Any metal other than iron and its alloys.(有色金属)

oxide Chemical compound containing oxygen and one other chemical element.(氧化物)

(the) **Peirce-Smith converter** A horizontal cylinder for converting with an opening that serves both for charging and discharging materials into the reactor as well as to collect emissions.(皮尔斯—史密斯转炉、皮氏卧式转炉)

(the) Peirce-Smith process A basic converting process for copper matte in the Peirce-Smith converter.(皮尔斯-史密斯吹炼)

pewter Any of numerous silver-gray alloys of tin with various amounts of antimony, copper, and sometimes lead, used widely for fine kitchen utensils and tableware.(白镴;锡镴制器皿)

platinum A chemical element with the symbol Pt and atomic number 78. It is a dense, malleable, ductile, highly unreactive, precious, silverish-white transition metal. Its name is derived from the Spanish term platino, meaning "little silver".(铂)

plutonium A radioactive chemical element with the symbol Pu and atomic number 94. It is an actinide metal of silvery-gray appearance that tarnishes when exposed to air, and forms a dull coating when oxidized.(钚)

potassium A chemical element with the symbol K and atomic number 19. Potassium is a silvery-white metal that is soft enough to be cut with a knife with little force. Potassium metal reacts rapidly with atmospheric oxygen to form flaky white potassium peroxide in only seconds of exposure.(钾)

precious metal Any of several metals, including gold and platinum, that have high economic value.(贵金属)

precipitation The process of separating a substance from a solution as a solid.(沉淀;析出)

pyrometallurgical process From pyrometallurgy, the branch of metallurgy involving processes performed at high temperatures, including sintering, roasting, smelting, casting, refining, alloying, and heat treatment. It means an ore-refining process dependent on the action of heat.(火法工艺)

reactivity The rate at which a chemical substance tends to undergo a chemical reaction.(反应性)

reductant A reducing agent which as it is oxidized is capable of bringing about the reduction of another substance.(还原剂)

refining The process of removing impurities (as from oil or metals or sugar etc.).(精炼)

rotary furnace A heat-treating furnace of circular construction which rotates the workpiece around the axis of the furnace during heat treatment; workpieces are transported through the furnace along a circular path.(回转炉;旋转加热炉)

rutile A lustrous red, reddish-brown, or black mineral, TiO_2, used as a gemstone, as an ore, and in paints and fillers.(金红石)

selenium A chemical element, atomic number 34, atomic weight 78.96, symbol Se.(硒)

silicate Chemical compound containing silicon, oxygen, and one or more metals, e.g., aluminum, barium, beryllium, calcium, iron, magnesium, manganese, potassium, sodium, or zirconium.(硅酸盐)

slag A nonmetallic product resulting from the interaction of flux and impurities in the smelting and refining of metals.(炉渣;矿渣)

smelting The process of extracting a metal from an ore by heating.(熔炼;提炼)

sodium A chemical element with the symbol Na and atomic number 11. It is a soft, silvery-white, highly reactive metal.(钠)

specific gravity The ratio of the mass of a solid or liquid to the mass of an equal volume of distilled water at 4℃ (39℉) or of a gas to an equal volume of air or hydrogen under prescribed conditions of temperature and pressure. Also called **relative density**.(比重)

sulfide Chemical compound containing sulfur and one other element or sulfur and a radical.(硫化物)

suspension smelting A category of direct smelting in which the reactions occur between gases and solids.(悬浮熔炼)

tantalum A metallic transition element, atomic number 73, atomic weight 180.9479, symbol Ta.(钽)

tellurium A chemical element, atomic number 52, atomic weight 127.60, symbol **Te**.(碲)

thorium a weakly radioactive metallic chemical element with the symbol Th and atomic number 90.(钍)

titanium A metallic transition element, atomic number 22, atomic weight 47.90, symbol Ti.(钛)

uranium A white, lustrous, radioactive, metallic element, isotopes of which are used in atomic and hydrogen bombs and as a fuel in nuclear reactors. Atomic weight 238.03, atomic number 92, symbol U.(铀)

volatility The quality of having a low boiling point or subliming temperature at ordinary pressure or, equivalently, of having a high vapor pressure at ordinary temperatures.(挥发性)

zirconium A metallic transition element, atomic number 40, atomic weight 91.22, symbol Zr.(锆)

zone refining Also known as zone purification, a technique to purify materials in which a narrow molten zone is moved slowly along the complete length of the specimen to bring about impurity segregation, and which depends on differences in composition of the liquid and solid in equilibrium.(区域精炼)

二、英语原文

Nonferrous Metals: Production and History
HY Sohn[①]

1 Nonferrous metals—all metals other than iron—occur in the earth's crust as chemical compounds (minerals) such as sulfides, oxides, silicates, and carbonates. Most important nonferrous metals occur as sulfides or oxides, often in the presence of iron. Some metals such as gold, however, may occur in the elemental form. Although large nuggets of gold have been found, most metals and minerals in nature are finely disseminated in gangue minerals such as silica or silicates. In certain parts of the earth, the concentrations of these metals or minerals are sufficiently high as to allow economical extraction of the metals. These concentrations have taken place owing to various geological processes. Major sources of metals and minerals can be found in the literature (e.g., Hayes, 1993; Dennis, 1965).

2 The first metal that was used by people was copper, which began to be smelted from the ore before 4000 BC in the Middle East. The first ores used were most likely relatively pure, oxidized copper minerals, such as malachite, that were found near the surface of the ground. They were probably smelted in a charcoal bed.

3 These easily found ores soon ran out, and minerals from deeper deposits containing sulfides of copper and other metals, especially arsenic, began to be used. Smelting of these arsenic-containing copper minerals produced the first bronze, thus bringing in the Bronze Age. Eventually, the toxic arsenic was replaced by tin to produce the more familiar copper-tin bronze, which has a much better combination of castability, ductility, workability, and hardness, and thus became the standard metal of use for the next thousand years. During the Bronze Age, gold and silver were also worked into ornaments, singly or to gild objects made of bronze.

4 Throughout the Bronze Age and the subsequent Iron Age, other metals were used. Lead was produced from its sulfide mineral, galena, by heating it on a charcoal bed. It was widely used for plumbing and pewter tableware by the Romans, which undoubtedly caused a great deal of lead poisoning. Lead was also used as a source of silver metal by separating the silver after oxidizing away the lead (cupellation). Mercury was extracted from its sulfide mineral cinnabar

---

[①] Professer of University of Utah, Salt Lake City, UT, USA.

from the time of ancient Greece and used for gilding bronze owing to its ability to form an amalgam with gold. During the first century BC in Rome, brass coins were made by adding about 28% zinc to copper. This was probably achieved by heating copper with a zinc carbonate mineral, because zinc metal itself was not isolated until the late medieval period.

5 During the Dark Ages, few noteworthy developments in metallurgy took place. With the advent of the Renaissance, three important books on metallurgy were written: *De La Pirotechnia* (Biringuccio, 1540), followed by *De Re Metallica* (Agricola, 1556), which was the standard reference until the eighteenth century, and *Berschreibung Allerfürnemisten Mineralischen Ertzt und Bergswecksarten* (Treatise on Ore and Assaying) (Ercker, 1574). These books described the early techniques for ore assaying, smelting, refining, and alloying, among other topics.

6 By the eighteenth century, the use of coke instead of coal as a reductant in metal production was developed. Later in the century, hydrogen was discovered, and soon its use in reducing metal oxides, such as tungsten oxide, to produce elemental metal was begun (Habashi, 1969).

7 The nineteenth century saw the discovery and isolation of many additional metals, owing in large part to the discovery of electricity and the development of electric cells and subsequently electric heating furnaces. Some of the important uses of electricity in nonferrous metal production during this period include the electrolytic refining of copper and the invention of the Hall-Heroult cell for producing aluminum. By the middle of the twentieth century, the foundations of most modern nonferrous metal production technologies were firmly established. For a recent summary on the subject as well as basic principles of nonferrous metals production, the reader is referred to a monograph by Sohn (2014a).

Production from Primary Sources

8 The production of metals from ores requires a series of physical and chemical separation steps. The ore is usually crushed and ground to liberate mineral grains from the gangue constituents. The mineral is then separated from the gangue by one of several physical methods taking advantage of the differences in specific gravity, magnetic property, or surface property. The most widely used concentration method is froth flotation, which is based on the difference in the affinity of solid surfaces toward air and water. In this process, gas (usually air) bubbles are injected from the bottom of a slurry that contains fine, solid particles. Certain solid particles attach themselves to the gas bubbles and rise to the top, leaving more hydrophilic particles behind. The concentrates thus obtained undergo chemical treatments to yield pure metals. Such a chemical treatment may proceed under a high temperature (pyrometallurgical process) or in a

low-temperature liquid (usually aqueous) solution. The extraction methods depend on a number of factors:

- physical and chemical properties of the minerals in the ore;
- physical and chemical properties of the metal to be produced;
- content of the metal in the ore;
- impurity content in the ore;
- properties of the associated gangue mineral;
- value of the metal to be produced; and
- value of by-products.

9 Sulfide minerals are most amenable to concentration by flotation, and the concentrate is usually treated by pyrometallurgical techniques with certain exceptions, as discussed below. Oxides are not easily concentrated by flotation, and thus they are directly leached or concentrated by other means such as gravity or magnetic separation.

10 Sulfide ores often contain significant amounts of precious metals in addition to other impurity elements. Pyrometallurgical processes allow the recovery of the precious metals because they are not easily oxidized and thus follow the metals to be produced. The low reactivities of the precious metals present an obstacle to their recovery during a hydrometallurgical treatment of sulfide minerals. Impurities such as bismuth, cobalt, molybdenum, selenium, tellurium, indium, cadmium, and germanium are recovered to an increasing extent (Sohn, 2014b).

11 The properties of the metal to be produced greatly affect the extraction method. Metals with comparatively low melting points ($\leqslant 1,550℃$) are amenable to pyrometallurgical smelting processes, during which the gangue constituents and easily oxidized impurities may be readily removed into a molten slag. The slag forms a separate liquid phase immiscible with the metalcontaining phase and can readily be removed. Metals with high melting points, however, are produced by first obtaining intermediate compounds that are either vapors or liquids at moderate temperatures. These intermediates are then purified by suitable means before being reduced to produce the metals—usually in the form of a sponge. Titanium, zirconium, and tantalum are examples of such metals.

12 Metals with low boiling points are often recovered as vapors. Zinc, for example, is recovered as a zinc vapor by reducing zinc oxide with carbon above the boiling point of zinc. If the vapor product contains impurities, they may be separated by fractional distillation (Sohn and Olivas-Martinez, 2014b).

13 The chemical reactivity of the metal to be produced is another important factor for metal extraction from the ore. Metals with relatively low reactivities, such as copper, nickel,

lead, and tin, may be recovered under relatively moderate conditions. If they occur as sulfides, they are extracted by reacting with oxygen in a molten phase. Their oxides can readily be reduced by carbon or hydrogen. These metals may also be recovered by leaching the mineral into aqueous solutions, which are purified and concentrated. The metal is then recovered from the solution usually by electrowinning but sometimes by precipitation.

14 The more reactive metals such as chromium and manganese are found in nature as oxides, which can be reduced by carbon but not hydrogen. The most reactive metals, such as titanium, zirconium, and aluminum, cannot readily be reduced to metal by carbon because they form carbides rather than metals. The production methods for these metals are discussed below.

15 The content of the metal in the ore affects the selection of the extraction method. Pyrometallurgical processing requires sufficiently high-grade concentrates. For example, a copper concentrate should contain more than about 25% copper to be suitable for smelting. Lower-grade materials that cannot be economically concentrated to this grade are collected in the form of a dump and leached in situ by circulating a leach solution through it over a long period of time. Metals that occur only in low-grade ores, such as gold and uranium, are generally recovered by leaching.

16 Major mineral sources for the most important nonferrous metals (copper, nickel, zinc, and lead) are sulfides. Pyrometallurgical production of metals from sulfides involves removing sulfur and iron that are usually present in the ore by oxidizing both with oxygen, sulfur to sulfur dioxide gas and iron to iron oxide (Sohn, 2014a,b). The sulfur dioxide is typically converted to sulfuric acid, and the iron oxide is absorbed into slag and discarded. In such a process, the condensed phase is in a molten state and the process is called "smelting". The smelting processes for copper and nickel are carried out in two major stages. The first is the mattemaking step, in which a large portion of the sulfur and most of the iron are oxidized, leaving a mixed sulfide of the metal and some iron. Such a mixed sulfide is called a "matte". In the mattemaking step, most impurities are removed into the slag. The matte is separated from the slag and undergoes further oxidation to remove the remaining iron and sulfur. This latter treatment is called "converting". The product of converting is a relatively pure metal that still contains some sulfur, oxygen, and other minor elements. Such a metal requires further purification at a high temperature (fire refining) and finally by electrolytic refining. In the electrorefining step, the crude metal is used as the anode and purified metal is deposited on the cathode. Impurities fall to the bottom of the electrolytic cell and are further treated to recover values, most important being the precious metals.

17 There are a number of different matte smelting and converting processes used in the

nonferrous metals industry (Sohn, 2014b; Sohn and Ramachandran, 1998). The most widely used matte smelting process is flash smelting. In this process, sulfide mineral concentrate, together with a slag component (typically silica), is injected with oxygen-enriched air into a furnace where the oxidation reactions take place. Since these reactions generate large amounts of heat, the process proceeds autogeneously once ignited. The molten matte and slag droplets fall to the bottom of the furnace (the settler), coalesce, and are separated into immiscible layers.

18 Matte smelting processes may in general be divided into two groups: suspension smelting of which flash smelting is an example, and bath smelting in which the sulfur and iron oxidation takes place in a molten bath. Another type of suspension smelting is the cyclone smelting process in which the gas stream, laden with the concentrate and flux particles, is injected into a cyclone reactor. The interaction between the gas and particles is more rapid and thus the process is more intense than in the flash smelting process, but this occurs at a high price of reactor material erosion and other operational difficulties.

19 In a bath smelting operation, the concentrate and flux are either charged on top of a molten bath or injected into the melt together with the process gas through injectors. Different types of injectors are used in different processes. The top submerged lance injects the stream below the surface of the molten bath, the top jetting lance is situated above the molten surface, the tuyere blows the stream into the side of the matte or the slag layer, and the gas-shrouded bottom injector delivers the process gas and solid particles vertically from the bottom of the furnace. In some cases, matte smelting is performed in an electric furnace.

20 The matte produced in the matte smelting step is separated from the slag and 'converted' into a crude metal in the converting process. The most important matte converting process is the Peirce-Smith process. In this process, molten matte is transferred from the matte-smelting furnace into the Peirce-Smith converter by the use of a ladle. This converter is a horizontal cylindrical vessel with an opening on top and a series of tuyeres along the side. Air is blown into the matte through these tuyeres, effecting the oxidation reactions. Although the Peirce-Smith process has been the industry standard since late nineteenth century, it has certain disadvantages with respect to the increasingly stringent industrial requirements in terms of atmospheric pollution, labor requirements, and productivity. The transfer and charging of molten matte and pouring of molten metal product provide opportunities for emission of sulfur dioxide and other harmful volatiles. Since the molten matte must be treated immediately, the operation of the Peirce-Smith converter is inherently coupled with that of the matte-smelting furnace. If one step is not functioning, the other must become idle.

21 A development to overcome these difficulties is the Kennecott-Outotec flash converting

process (Sohn, 2014b; Sohn and Ramachandran, 1998). In this process, the molten matte produced in the matte smelting step is granulated, ground into fine particles, and stockpiled. These matte particles are converted in a flash furnace that is similar to a flash smelting furnace but smaller. The chief advantages of this converting process include a continuous and closed operation, decoupling of the matte smelting and converting steps, and generation of a continuous, rich sulfur dioxide stream that facilitates sulfuric acid production. The sulfur dioxide emission is greatly reduced in this process, resulting in an overall sulfur capture of more than 99.9% compared with a wide range of overall sulfur capture of 80%-99% in various other processes.

22 Other converting processes are carried out in a molten bath equipped with top submerging or jetting lances, similar to some of the matte smelting operations discussed above.

23 Zinc and lead sulfides are usually oxidized first to oxides in a process called roasting, followed by reduction of the oxides by the use of carbon (coke or coal) as the reducing agent. The roasted oxides are charged as solids into various types of reactors, depending on the process used in which the reduction reaction takes place (Sohn and Olivas-Martinez, 2014a, b). An exception to this two-step approach is the commercialized QSL lead production process. In this process, the oxidation and reduction reactions are carried out at the opposite ends of a long, horizontal cylindrical vessel, the lead oxide remaining as a liquid. The oxidation is effected by oxygen injected through gas-shrouded bottom injectors, and the reduction is done by the injection of an oxygen-coal mixture through similar bottom injectors that generates a high-temperature reducing atmosphere (Sohn and Olivas-Martinez, 2014a; Sohn and Ramachandran, 1998).

24 Hydrometallurgical production of nonferrous metals involves leaching metal values from concentrates or ores into aqueous solutions. The solutions are purified, concentrated, and separated into streams containing different elements. The metals are recovered from these solutions by electrolysis (electrowinning) or precipitation. For copper production, hydrometallurgical processes are mainly applied to oxidized or partially oxidized ores and low-grade dump stocks. While pyrometallurgical processes involve relatively simple and fast chemical reactions, hydrometallurgical processes generally involve complex and slow reactions that require large volumes of solutions. Other problems include the difficulty of recovering precious meals, which are not leached, and the generation of dilute solutions containing sulfates, which require treatments before disposal. Approximately 15% of world copper production is by hydrometallurgical methods (Habashi, 1997). Nickel production involves more hydrometallurgical steps, especially when the ore contains a substantial amount of copper. The lixiviants are typically sulfate, chloride, or ammonia media (Habashi, 1997). A large portion of zinc production

worldwide is by hydrometallurgical treatment. Zinc sulfide minerals are roasted to oxides, which are leached into an acid media. The solution is then purified before zinc is recovered by electrowinning.

25 Despite the fact that there exist a number of significant hydrometallurgical operations for nonferrous metal production, most attempts to develop cost-effective hydrometallurgical processes to replace large-scale pyrometallurgical ones have met with failure. The inherent efficiencies of smelting make it difficult to find a better alternative. These efficiencies stem from the fact that in smelting the raw concentrate is used as a valuable fuel to provide the process heat using only oxygen from the air, the process takes place rapidly at a high temperature, and nearly 100% of metal values is recovered, all in a relatively few process steps. The nearly complete metal recovery includes that of precious metals and other by-products, which is much more difficult in hydrometallurgical processes. The disadvantages of pyrometallurgy include difficulties in containing molten materials, severe conditions that require close control and skilled technical operators, and large inventories of metal in the refinery (Sohn, 2000).

26 Important nonferrous metals that are extracted from nonsulfide ores include light metals (e.g., aluminum, titanium, and magnesium), precious metals (e.g., gold, silver, and platinum group metals), refractory metals (e.g., tungsten, molybdenum, niobium, and zirconium), alkali metals (e.g., lithium, sodium, and potassium), and radioactive metals (e.g., uranium, plutonium, and thorium). The extraction methods for these metals vary as widely as their occurrence and properties, and the reader is referred to other articles in this encyclopedia and elsewhere (Sohn, 2014a; Habashi, 1997; Gill, 1980; Gilchrist, 1980; Ryan, 1968) for further details. Here, only the primary production methods for selected metals in this category will be discussed (Sohn, 2014a).

27 Aluminum, like most reactive metals, occurs in nature in some oxidized form. Aluminum-containing minerals are usually quite stable chemically, and thus are difficult to reduce to the elemental form. The only important ore for aluminum production is bauxite, a mixture of hydrated aluminum oxides ($Al_2O_3 \cdot H_2O$ and $Al_2O_3 \cdot 3H_2O$), which usually contains iron oxide and silica as impurities. Alumina is produced from bauxite by the Bayer process, which involves leaching with caustic solution to dissolve aluminum oxide as $NaAlO_2$ and its subsequent hydrolysis and precipitation of $Al(OH)_3$. The latter is calcined to alumina, Al_2O_3, which is reduced electrolytically to produce aluminum metal. The electrolysis is done in a Hall-Heroult cell containing a molten mixture of cryolite (Na_3AlF_6), AlF_3, and CaF_2 in which alumina is dissolved. The electrodes are a consumable carbon anode and a carbon cathode that forms the lining of the cell. The reduced aluminum metal collects at the bottom of the cell.

28   Another important light (reactive) metal is titanium. The major ores of titanium are rutile and ilmenite. The naturally occurring rutile is mainly TiO_2(49.5%) with small amounts of iron oxide, silica, and other metal oxides. Ilmenite ore is mainly $FeTiO_3$ with 35%–60% TiO_2 content. It is treated to increase the TiO_2 content by either electric smelting or other chemical methods. The largest use (about 95%) of these titanium minerals is for the production of TiO_2 pigments. For the production of titanium metal, the TiO_2-containing raw material is chlorinated in a fluidized bed in the presence of coke to first produce $TiCl_4$, which is purified by distillation. The most widely used method for producing titanium metal from $TiCl_4$ is the Kroll process. In this process, $TiCl_4$ is slowly added to a reactor that contains molten magnesium to form titanium sponge and $MgCl_2$. After the $MgCl_2$ and excess magnesium are distilled away, the titanium sponge is consolidated by vacuum arc melting. The Kroll process is also used to produce zirconium from zircon ($ZrSiO_4$).

29   The production of precious metals deserves some mention. Gold is largely produced by dissolving it with sodium cyanide from the powdered ore into the solution as $Na[Au(CN)_2]$. After filtration, gold is recovered from the solution by adding zinc dust, in which case zinc is dissolved into the solution and gold is precipitated, or by adsorption onto activated carbon. In the latter case, the loaded carbon is eluted with a caustic cyanide solution to redissolve the gold, which is recovered by electrowinning. Gold is also produced in large quantities as a by-product of copper smelting. Gold contained in the copper concentrate follows copper through the entire smelting process. It is separated in the copper electrorefining step by falling to the bottom of the electrolysis cell as a slime, together with other minor elements.

30   Silver is produced by several different processes depending on the chemical nature of the mineral in which it occurs (Habashi, 1997). The cyanide process described above for gold production is the dominant method for silver production from true silver ores. Large quantities of silver are also produced from lead and lead-zinc ores as well as from copper and nickel smelting via the anode slime.

Production from Secondary Sources

31   Like iron and steel, nonferrous metals are also produced from secondary sources. The main secondary source is scrap metal, which is generated during the primary production or the fabrication of objects as well as from recycled items. The aluminum industry has been very successful in using recycled metal (mostly used beverage cans) to produce primary aluminum. Secondary lead is mainly produced by the recycling of waste lead-acid batteries and amounts to about 75 % of the total lead produced (B 4 million tons as of 2010) worldwide excluding China

(Hassall and Roberts, 2010). These recycled materials can be processed by primary smelters using the lead blast furnace or the QSL process or by secondary smelters using the rotary or reverberatory furnaces (Sohn and Olivas-Martinez, 2014a). The recycled batteries are first separated into metallics (Pb) and a paste containing lead sulfate and oxides. In primary smelters, this paste is mixed with the sulfide concentrate charge. Secondary smelters reduce the paste to metallic lead by carbon. Gold is recovered in large quantities from used gold alloys containing other metals such as copper, nickel, and silver, from sweeps generated in the noble metal processing industry, and from used gold-containing coatings generated largely by the electronics industry (Habashi, 1997). Secondary silver is produced from sources similar to those of secondary gold, plus recycled photographic materials.

32 With the richer ores being depleted and the grade of the primary ore mined decreasing, the costs of producing concentrate and thus of primary metal production will continue to increase. Further, taking into consideration the increasing costs for more stringent environmental protection as well as the ever-increasing costs of energy, the recycling of used metals and the improvement of product yield during the primary production will become increasingly important.

Value Addition and Refining Processes

33 Up to this point, the primary production of nonferrous metals has been discussed, i.e., the production of individual metals with the highest purity that is required for the widest applications. Some of these applications require much higher purity, and some others require a mixture of metals in various proportions. For example, certain applications of copper in the electronic industry require greater than 99.999% purity with less than 0.001% oxygen content, compared with 99.97%–99.99% purity with 0.01%–0.03% oxygen for copper mass produced by the typical electrorefining step. Alloying is often necessary to produce the right properties for certain applications, such as wear and corrosion resistance, hardness, toughness, high-temperature strength, color, and electric and magnetic properties.

34 With the increasing demands for newer and more advanced materials based on ever-increasing technological developments, value addition by modifying the properties of existing materials and, indeed, by discovering new materials with new properties has become an important aspect of the nonferrous metal production industry. Examples can be found in applications that include semiconductors, ultrapure metals, chemical-vapor-deposited metallic films, intermetallics, and composites (Sohn, 1991).

35 To produce metallic materials with stringent property specifications requires mixtures with extremely narrow composition ranges and extremely low impurity contents. It is obvious

then that the starting materials must be of sufficiently high purity. Metal refining and purification methods can be achieved by either a chemical method or a physical method.

36 The removal of impurities by forming chemical compounds takes advantage of the fact that different metals have different affinities to oxygen, sulfur, or other nonmetallic elements. Such compounds may be in the form of a solid, liquid, or gas possessing different physical properties from the metal, which can be used in their separation.

37 Compound formation is usually carried out in the molten phase because of the following factors:

• The contact between the metal and the compound-forming agent can be achieved readily in the liquid phase.

• Subsequent to refining, alloy-forming elements can conveniently be added and mixed uniformly.

• The molten product can easily be formed into various end shapes.

38 An excellent example of metal refining and purification by compound formation is the production of high-purity lead, which consists of several chemical treatments of crude lead produced in the primary smelting operation (e.g., Sohn and Olivas-Martinez, 2014a; Hayes, 1993; Habashi, 1997).

39 Some chemical refining methods involve forming a volatile compound of the metal to be refined, separating it from the crude metal, and decomposing it to obtain the purified metal. The decomposition step typically yields the gaseous reactant that is recycled to the initial step of volatile formation. Examples are the Mond (carbonyl) process for nickel purification and the iodide processes for the refining of such metals as zirconium, titanium, and thorium (Habashi, 1986).

40 Vacuum refining makes use of the different volatility of different elements and compounds. Although most widely used in steel refining to decrease hydrogen and oxygen content to low levels, vacuum refining also finds application in purifying nonferrous metals (Yongnian and Rixin, 1994).

41 Zone refining is used to prepare very high-purity metals, usually from metals that already have relatively high purity. This technique takes advantage of the difference in the solubility of an impurity in the solid and liquid phases (Pitt, 1979; Hayes, 1993; Habashi, 1986). In this process, a bar or a rod of a starting material is held vertically or horizontally in an inert environment. A narrow zone of molten phase is formed at one end by some means, such as an induction heater or electron beam (Yu et al., 1992; Habashi, 1986) and is moved along the length of the specimen. The impurity, which typically has a greater solubility in the liquid

phase, is concentrated in the moving molten zone and is thus carried toward the end of the specimen. Making multiple passes of the molten zone in the same direction is typically required to produce a sufficiently high purity. Predominantly applied to the production of semiconductor materials, almost any metal can be refined by this process.

42 Electric arc melting is used for refining as well as casting reactive metals with high melting points such as titanium, tantalum, zirconium, and molybdenum. Under the high temperature generated by the electric arc, impurities are removed by volatilization. For obvious reasons, this process is conducted in a high vacuum or an inert atmosphere. The electrode may be made of an inert material or of the metal to be refined. In the latter case, the electrode melts, undergoes refining, and forms an ingot in a watercooled crucible which acts as the other electrode. Electroslag refining can be included in this category, except that a slag pool is maintained above the refined ingot to effect the refining process.

43 Metals that have relatively low boiling points such as zinc and mercury are purified by simple distillation when there is a large difference in the boiling points of the constituents, and fractional distillation when the difference in boiling points is small. Fractional distillation is essentially repeating the evaporation and condensation as many times as required within a single column with the vapor and the liquid phase traveling in opposite directions. Thus, the metal with a lower boiling point is concentrated into the vapor phase and vice versa.

Wastes and Environmental Issues

44 As in any other industry, production of metals creates potential pollutants. In nonferrous metal production from sulfide minerals, large amounts of sulfur dioxide are generated, as discussed above. The method of fixing this gas is to convert it to sulfuric acid. In many parts of the world, sulfur dioxide emission into the atmosphere is still a serious problem. The situation, however, is rapidly changing owing to the enactment of more strict regulations and, as a result, of development and adoption of more advanced smelting technologies. In fact, most of the new technologies developed and adopted by the nonferrous metals industry have been largely driven by environmental concerns. The most remarkable improvement has been the rapid increase in sulfur capture. Older technologies that provided opportunities to generate sulfur dioxide gas as fugitive or low-concentration process streams continue to be replaced by technologies that produce streams with higher concentrations and at uniform rates (Sohn, 2014a,b; Sohn and Ramachandran, 1998). This enables a higher degree of sulfur capture in the acid plant. A further consequence of this development is the improved conditions for fixing sulfur as the element rather than sulfuric acid. Sulfuric acid has many useful applications and thus is an

important chemical agent. Being a corrosive and hazardous material, however, it is difficult to store and transport. When produced by a smelter in larger quantities it seldom has an attractive market readily available. The capture of sulfur dioxide as elemental sulfur thus has many advantages because sulfur is a solid and inert at room temperature. It can readily be stored, disposed of, or converted to other useful products including sulfuric acid when and where needed. Continued effort is being devoted to the development of methods for economically converting sulfur oxide to elemental sulfur (Kim and Sohn, 1999; Kim, 1999).

45 The mineral concentration processes generate tailings consisting of gangue minerals in the ore. They are either returned to the mine or deposited in ponds. Slags from smelting processes contain up to 1% copper, often less. These slags are slow-cooled, ground, and treated by flotation to recover the copper value as a slag concentrate, which is recycled to the smelting process. In some cases, the smelter slag is treated in an electric furnace to recover the copper as a matte. The remaining slag is disposed of as waste slag. The converter slags may contain up to 10% copper, and in the case of the Kennecott-Outotec flash converting process about 18% copper. In the latter case, the slag is granulated, ground, and mixed into the concentrate feed charged to the flash smelting furnace. The cost for cleaning the smelter slag can amount to 15% of the total copper smelting cost (Sohn and Ramachandran, 1998). Improvements in decreasing the copper content in the slag during the smelting process and in increasing the efficiency of copper recovery from the slag remain an important technical challenge.

46 The hot gases from the smelting and converting furnaces contain significant heat values that are recovered in waste heat boilers. When the dust-laden gas is sufficiently cooled by passing it through the waste heat boiler, the dust is collected typically by an electrostatic precipitator. The dusts contain metal values as well as impurities and are returned to the smelting furnace.

47 Hydrometallurgical processes have been considered as an alternative to pyrometallurgical ones, especially from the pollution point of view. The problems of minimizing gaseous and fume emissions are, however, replaced by those of the disposal of leaching residues and large amounts of waste water that contain acids and other contaminants. Solid wastes from pyrometallurgical operations are slags, which are relatively innocuous and can easily be disposed of on a confined area of land. In comparison, it is much more difficult to contain liquid wastes (Sohn, 2000).

48 A common perception has existed that the primary metal production industry involves dirty plants and unsophisticated technologies. This is far from being true, especially for modern plants. Many of the steps in metal production are controlled by modern instruments and computers in a clean environment, and the process vessels have become much more enclosed

with continuous transfer of materials through confined conduits.

49 The production of individual metals and the details of various metal extraction techniques are described in other appropriate articles in this encyclopedia.

References

Agricola, G., 1556. De Re Metallica, (H.C. Hoover, L.H. Hoover, Trans.). Basel: Froben. 1950. New York: Dover.

Biringuccio, V., 1540. De La Pirotechnia, (C.S. Smith, M.T. Gnudi, Trans.). Venice; 1966. Cambridge, MA: MIT Press.

Dennis, W.H., 1965. Extractive Metallurgy—Principles and Application. London: Pitman.

Ercker, L., 1574. Berschreibung Allerfürnemisten Mineralischen Ertzt und Bergswecksarten (Treatise on Ore and Assaying), (A.G. Sisco, C.S. Smith, Trans.). Prague; 1951. Chicago, IL: University of Chicago Press.

Gilchrist, J.D., 1980. Extraction Metallurgy, second ed. Oxford: Pergamon.

Gill, C.B., 1980. Nonferrous Extractive Metallurgy. New York: Wiley.

Habashi, F., 1969. Principles of Extractive Metallurgy: Vol.1. General Principles. New York: Gordon and Breach.

Habashi, F., 1986. Principles of Extractive Metallurgy: Vol.3. Pyrometallurgy. New York: Gordon and Breach.

Habashi, F., 1997. Handbook of Extractive Metallurgy, vols. II-IV. Weinheim, Germany: Wiley-VCH.

Hassall, C., Roberts, H., 2010. In: Siegmund, A., et al. (Eds.), Lead-Zinc 2010. Hoboken, NJ: John Wiley & Sons, Inc., pp. 17-25.

Hayes, P.C., 1993. Process Principles in Minerals and Materials Production. Sherwood, QLD, Australia: Hayes Publishing.

Kim, B.-S., 1999. Reduction of sulfur dioxide to elemental sulfur by cyclic process involving calcium sulfide and sulfate. University of Utah. PhD Thesis.

Kim, B.-S., Sohn, H.Y., 1999. The reduction of sulfur dioxide by calcium sulfide to produce elemental sulfur. In: Mishra, B. (Ed.), EPD Congress 1999. Warrendale, PA: TMS, pp. 131-188.

Pitt, C.H., 1979. Zone refining. In: Sohn, H.Y., Wadsworth, M.E. (Eds.), Rate Processes of Extractive Metallurgy. New York: Plenum, pp. 408-419.

Ryan, W., 1968. Non-Ferrous Extractive Metallurgy in the United Kingdom. London: Institution of Mining and Metallurgy.

Sohn, H.Y., 1991. The coming-of-age of process engineering in extractive metallurgy—The 1990 extractive metallurgy lecture. Metall. Trans. B 22B, 727-754.

Sohn, H.Y., 2000. Nonferrous metals production—Advances in process technology and environmental protection. In: Mishra, B., Yamauchi, C. (Eds.), Second International Conference on Processing Materials for Properties. Warrendale, PA: TMS, pp. 3-12. 6 Nonferrous Metals: Production and History.

Sohn, H.Y., 2014a. Chapter 2 Non-Ferrous process principles and production technologies, Vol. 3. In: Seetharaman, S. (Ed.), Industrial Processes Part A, Monograph within a super treatise on Treatise on Process Metallurgy. Oxford, UK and Waltham, MA, USA: Elsevier.

Sohn, H.Y., 2014b. Industrial Technologies for Copper Production, Treatise on Process Metallurgy, Vol. 3. Industrial Processes Part A. Oxford, UK and Waltham, MA, USA: Elsevier. Section 2.1.2, pp. 591-600.

Sohn, H.Y., Olivas-Martinez, M., 2014a. Lead Production, Treatise on Process Metallurgy, Vol. 3. Industrial Processes Part A. Oxford, UK and Waltham, MA, USA: Elsevier. Section 2.3.1, pp. 671-693.

Sohn, H.Y., Olivas-Martinez, M., 2014b. Zinc Production, Chapter XI-3.2 Treatise on Process Metallurgy, Vol.3. Industrial Processes Part A. Oxford, UK and Waltham, MA, USA: Elsevier. Section 2.3.2, pp. 693-700.

Sohn, H.Y., Ramachandran, V., 1998. Advances in sulfide smelting—Technology, R & D, and education. In: Asteljoki, J.A., Stephens, R.L. (Eds.), Sulfide Smelting '98, Current and Future Practices. Warrendale, PA: TMS, pp. 3-37.

Yongnian, D., Rixin, L., 1994. Vacuum metallurgy—An important direction in reforming conventional nonferrous metallurgy. In: Sohn, H.Y. (Ed.), Metallurgical Processes for Early Twenty-First Century. Warrendale, PA: TMS, pp. 421-427.

Yu, H.S., Kim, J.S., Rhee, K.-I., Lee, J.-C., Sohn, H.Y., 1992. Production of ultrapure tungsten by solvent extraction with a D2EHPA/TOPO mixture and electron-beam zone refining. Int. J. Refract. Met. Hard Mater. 11, pp. 317-324.

三、分析与讲解

有色金属的重要性，每个人都非常清楚。买电线时我们知道里面必须是铜芯，焊接线路连接电子元器件时我们需要焊锡，装修房屋时我们经常购买铝合金门窗，建造飞机时我们需要用到钛，而手机里面用的是锂电池……

有色金属是相对黑色金属而言的。黑色金属的英文为 ferrous metals，直译为铁类金属，是铁、锰、铬三种金属的统称；而有色金属的英文 nonferrous metals 直译为非铁类金属，即除铁、锰、铬以外的金属的统称。

《大辞海（材料科学卷）》金属材料——有色金属材料一节对"有色金属"的定义为："元素周期表中除铁、铬、锰三种金属以外的所有金属元素的统称。可分四类：（1）重金属，包括铜、铅、锌、镍等；（2）轻金属，包括铝、镁、钛等；（3）贵金属，包括金、银、铂等；（4）稀有金属，包括钨、钼、钽、铌、钍、铍、铟、锗、稀土金属等。由于稀有金属在现代工业中具有重要地位，故常从有色金属中划出来单独成为一类。"也有人提出有色金属应该按表 10.1 所示来划分：

表 10.1 有色金属划分表

有色金属分类									
有色轻金属（8 种）	钠	钾	镁	钙	锶	钡	钛	铝	
有色重金属（10 种）	钴	镍	铜	锌	镉	汞	锡	铅	锑	铋
贵金属（8 种）	钌	锇	铑	铱	钯	铂	银	金	
稀有轻金属（4 种）	锂	铷	铯	铍					
稀有难熔金属（8 种）	锆	铪	钒	铌	钽	钼	钨	铼	
稀有分散金属（4 种）	镓	铟	铊	锗					
稀有稀土金属（17 种） 轻稀土金属（7 种）	镧	铈	镨	钕	钷	钐	铕		
稀有稀土金属（17 种） 重稀土金属（10 种）	钪	钇	钆	铽	镝	钬	铒	铥	镱	镥
准金属（5 种）	硼	硅	砷	硒	碲				
稀有放射性金属（23 种）	钫	镭	锕	钍	镤	铀	镎	钚	镅	锔
稀有放射性金属（23 种）	锫	锎	锿	镄	钔	锘	铹			
稀有放射性金属（23 种）	𬬻			𬭳	𬭛	𬭶				
稀有放射性金属（23 种）	钅卜	钅立								

我国是有色金属生产和消费大国。根据中国有色金属工业协会的报告，2022 年，我国十种常用有色金属产量为 6793.6 万吨，其中，精炼铜产量为 1106.3 万吨，原铝产量为 4021.4 万吨，氧化铝产量 8186.2 万吨，而 2022 年全球精炼铜、原铝、氧化铝分别为 2508.48 万吨、6830.46 万吨、1.4 亿吨。与此同时，2022 年我国精炼铜消费量为 1415 万吨，原铝消费量为 3985 万吨。这些数字反映了我国作为制造业大国的雄厚实力。

本文简要介绍了常用有色金属的生产。文章难度一般，但翻译时陷阱不少，详见如下文字：

作者姓名：这篇文章的作者英文名为 H. Y. Sohn，是美国犹他大学的教授。登录犹他大学的官网，发现该教授英文全名为 Hong Yong Sohn，是韩国人，毕业于首尔国立大学；再查中、英、韩姓名对照表，发现其中文名可以译为"孙洪龙"。

科技翻译时也常会遇到专有名词翻译的问题。一般说来，专有名词翻译有四大原则：(1)约定俗成原则，例如 the White House 不译成"白房子"而是"白宫"。(2)名从主人原则，即译音要尽量接近原文发音，如日喀则不是 Rikaze，而是 Shigatze。(3)专名专译原则，即同名同译。如在中国大陆 Donald Trump 普遍被译为"特朗普"，就不能译成中国台湾等地所用的"川普"。(4)音译为主原则，但音译日本人、韩国人的名字时需要特别注意，比如这里的 Hong Yong Sohn 就不是"洪勇孙"或"洪永颂"。

第 1 段：(1) all metals other than iron 是插入语。一般说来，插入语都可以翻译成插入语，这里也适用，但考虑到这个插入语事实上是对有色金属下定义，非常重要，所以应该译成主句的一部分。(2) in the elemental form 意思很清楚："别的金属都以化合物的形式存在，但金元素可能聚集在一起成为金块，不含任何别的元素"，但表达起来不易。这里为了强调，最好添加上"不含杂质的"。(3) finely disseminated 虽然意思简单，但同样不好译，不能简单地说"细细地分散在"，也不能说"以细小的颗粒形式"，因为有些金属太细(too fine)，不能被称为颗粒，所以比较恰当的译法是"以细小的粉粒散布在"。(4)最后两句都出现了 concentrations 这个词，根据上下文，翻译时进行灵活处理是非常必要的，前者可以译为"含量"，后者可以译为"富集"。

第 2 段：(1) that was used by people 是被动语态。被动语态在科技英语中比较常用，翻译成汉语的被动句或主动句都可以，比如这里既可以译为"人们最早使用的金属是铜"，也可以说"最早为人们使用的金属是铜"。(2) which began to be smelted... 是定语从句，从句中 which 是关系代词。汉语中没有关系代词，所以英语定语从句的关系代词要么译成名词，要么译成指示代词。

第 3 段：(1)虽然人们说英语形合，汉语意合，但有时汉语句子也需要衔接的词语，原文衔接用的是 and，这里须根据前后句关系将其译成"因此"。(2) the Bronze Age 译成"青铜时代"或"青铜器时代"肯定没问题，前面加上"人类历史上的"，意思更清楚。(3) Eventually 本意是"最终"，但这里译成"最终"却感觉不太好，因为最终发生的是青铜器时代为铁器时代取代，而不是铜锡青铜代替了含砷的青铜，所以改译成"后来"更恰当。(4) copper-tin bronze 是科技术语，一般译成"锡青铜"，无须画蛇添足译成"铜锡青铜"。科技术语的翻译以标准化为目标，不可闭门造车。

第 4 段：other metals were used 可以译为"人们使用其他金属"，但考虑上下文语气的连贯，最好在"使用"前加"也"，因为同样的原因，翻译 which undoubtedly caused a great deal of lead poisoning 时，最好在前面加上"结果"两个字。

第 5 段：（1）*De La Pirotechnia*，*De Re Metallica*，*Berschreibung Allerfürnemisten Mineralischen Ertzt und Bergswecksarten* 是三本书名，属专有名词。专有名词的翻译需要特别小心：如果已经有了约定俗成的译法，即使该译法不那么准确，也要采用该译法。如果没有约定俗称的译法，那么可以自行翻译，但最好用括号带上原专名。（2）which was the standard reference until the eighteenth century 可以译成前置修饰语，提出句子，作为补充信息，放在三本书名之后，更利于表达的流畅。

第 6 段：（1）was developed 译成"被开发"，意思不够完整，译成"被成功开发"就好多了。（2）"例如氧化钨"（such as tungsten oxide）可以直接放在"金属氧化物"后，放在括号中再放在"金属氧化物"之后，语句显得更流畅。

第 7 段：（1）subject 有"学科""话题"不同意思，考虑到这里的语境，译成"话题"应该更贴切。（2）考虑到 Sohn 就是作者本人，所以应将 a monograph by Sohn 译为"作者本人的专著"，并用括号给出参考文献信息。

第 7、8 段间的小标题：原小标题是 Production from Primary Sources，但翻译时需要增加"有色金属"，否则表达就很别扭。primary sources 事实上指"矿石"，也即"初始原料"。

第 8 段：（1）翻译 The ore is usually crushed and ground… 时最好加上"首先"两个字，同后面的 then 呼应。这一句采用了被动语态，但用主动语态意思也清楚（"矿石首先需要破碎、研磨"），只不过这里用了"需要"，前面一句的 requires 最好就不译成"需要"而译成"离不开"，这算翻译方法中的"正说反译"。（2）leaving more hydrophilic particles behind 字面意思是"将更多的亲水颗粒留在后面"，考虑到前后句，"留在后面"就是"留在泥浆底部"。（3）翻译…may proceed under a high temperature (pyrometallurgical process) or in a low-temperature liquid (usually aqueous) solution 需要特别注意，这里的 a high temperature 并不修饰 solution，不能译成"在高温或低温液体溶液中进行"，而需要分成两个子句表达："可能是在高温下进行的（火法冶金），也可能是在低温液体（通常是水）溶液中进行"。（3）"物理和化学性质"（physical and chemical properties）可以简化表达为"理化性质"。（4）最后的 and 译成"以及"，原文前面的分号既可以译成分号，也可以译成逗号，甚至干脆去掉。中英文有着略微不同的标点符合系统，因此英汉互译时不能完全照抄原文的标点符号，偶尔需要适当改变，比如将英文的逗号译成中文的顿号，或者将中文的逗号译成英文的句号。

第 10 段：（1）significant amounts of 意思是"大量的"，译成"丰富的"，更贴切。（2）low reactivities 说的是贵金属反应性低，译成"性质不活泼"，意思不变，表达更符合汉语规范。（3）follow the metals 中的 metals 显然指贵金属以外的金属。（4）原文最后三句话之间关系紧密，加上适当的连接词，合译成一句话，更显得一气呵成。

第 11 段：（1）to be produced 是"待生产的"，也就是"计划生产的"或"准备生产的"。

(2) usually in the form of a sponge 译成前置修饰语更好，当然放在后面也可以（"……再还原生产出金属，通常呈海绵状"），但不够紧凑。

第 12 段：(1) 原文 Zinc 前面是逗号，但考虑到第二句和第一句的亲密关系，翻译成中文时应改成逗号。(2) is recovered as a zinc vapor 译成"以气态锌的方式被回收"意思没问题，但表达别扭，译成"得到的是气态锌"就好多了。

第 13 段：(1) 可以先译 metal extraction from the ore，这样一来前面的 to be produced 就无须翻译了。(2) 第三、四两句说的是两种不同的情形，翻译时最好采用同样的句型，形成比照。(3) 最后两句可以译成一个汉语句子，但这样的话，译句就比较长，需要注意使用表示先后次序的词语，如"然后""最后"。当然，也可以译为"回收这些金属，需要先将矿物浸出到水溶液，提纯、浓缩，再从水溶液中通过电解沉积或沉淀提取，其中电解沉积法常用，沉淀法偶用"。

第 14 段：虽然第二句是 The most reactive metals... cannot readily be reduced...，但从上下文可以看出，这里的 metals 事实上指的是这些金属氧化物矿，否则就没有还原的必要，所以翻译时需要加上"氧化物"。

第 15 段：(1) 第二句中的 smelting 译成"冶炼"也可以，不过前面说的是用火法工艺，所以译成"熔炼"更贴切。(2) are collected in the form of a dump 字面是"以矿堆的方式收集"，其实就是"收集起来然后堆成一堆"。(3) circulating a leach solution through it 是"让浸液循环通过矿堆"，既然是液体，改"通过"为"流过"更好。

第 16 段：(1) 因为前面提到了几种有色金属，所以 are sulfides 翻译时需要加一个"都"（"都是硫化物"）。(2) matte 的意思是"锍"，是铜、镍等有色金属冶炼过程中产生的中间产品，为各种金属硫化物的互溶体。意思很清楚，但 the mattemaking step 却不好译。考虑到这是一个专业术语，查找相关资料，就能确定其可以译为"造锍熔炼过程"。(3) a large portion of 和 most of 意思相近，合并译为"大部分"。

第 17 段：(1) There be 句型在英语中很常见，翻译时根据情况有多种处理方式，这里是用原句中的表语（matte smelting and converting processes）作主语。(2) 英文中喜欢使用代词，中文喜欢重复名词，In this process 可以译成"在这个过程中"，但译成"闪速熔炼时"更符合汉语表达习惯。(3) 既然是闪速熔炼，那么 a furnace 当然是指"闪速炉"；where 引导定语从句在这里无须翻译，意思已经很清楚了。

第 18 段：(1) 考虑到 suspension smelting 和 bath smelting 关系紧密，先翻译这两个术语，然后再翻译术语后的定语从句，这样表达比较好。(2) the process is more intense 中的 the process 指"熔炼过程"，也就是"反应过程"。

第 19 段：(1) The top submerged lance, the top jetting lance, the tuyere, the gas-shrouded bottom injector 分别指四种不同的喷射器或者说喷枪，所以应译成结构类似的词

组。同样，四个术语所在句子的译句也应结构类似。(2)第三句中的 stream 指的是高速运动的炉料像气流一样，考虑到"炉料流"听起来别扭，所以直接译成"炉料"。

第20段：(1)This converter is a horizontal cylindrical vessel 不能直译，否则读来别扭，可以意译为"皮尔斯—史密斯转炉为卧式，呈圆筒状"。(2)with respect to the increasingly stringent industrial requirements 字面意思是"鉴于日益严格的工业标准"，考虑前后句，不妨译为"但在如今对工业生产要求日益严格的背景下"。(3)最后一句所谓 is inherently coupled with，就是指两种炉子的运行要很好地协调，因为一个出问题，另一个就无法运行。

第21段：(1)第三句中的 matte，因为已经制成颗粒了，不再是熔体了，所以就不能译成"熔锍"，吹炼的有色金属可以是铜，也可以是镍等，所以译成"金属锍"。同一句中的 flash furnace，既然是用来吹炼的，为了与 flash smelting furnace 区分开来，不妨译为"闪速吹炼炉"。(3)an overall sulfur capture of more than 99.9% compared with a wide range of overall sulfur capture of 80%-99% in various other processes 也可以译成"与其他工艺80%—99%的总体硫捕获率比，(闪速吹炼工艺的)总体硫捕获率超过99.9%"，但不如"总体硫的捕获超过99.9%，相比之下，采用其他工艺，总体硫捕获率在80%至99%之间"这样的表达更简洁明快。

第23段：(1)第一句中的 process，考虑到具体语境，译成"环节"更恰当。(2)The roasted oxides 字面意思是"焙烧过的氧化物"，但从前一句可以得知，"焙烧"的是硫化物，得到的是氧化物，所以应译成"焙烧得到的氧化物"。(3)An exception to this two-step approach is the commercialized QSL lead production process 字面意思是"商业化的 QSL 铅生产工艺是这种两步法工艺的例外"，翻译起来就是"两步法之外，还有一种商业化的 QSL 铅生产工艺"。

第24段：(1)原文前三句说的是湿法冶金工艺如何提取金属，翻译时可合并成一句话。同样，最后三句都是谈如何使用湿法冶金提取锌，也可以合并成一句译出。(2)For copper production 译成"生产铜时"肯定没问题，但读来有头重脚轻的感觉，改译成"生产铜金属时"，效果更好。(3)which are not leached 虽然是定语从句，但起到的是原因状语的作用，所以可以译为原因状语"因为贵金属无法浸出"。

第25段：(1)第一句中的 operations 译成"操作"，不太符合语境；译成"工艺"，过于概括，不妨译为"技术"。(2)The inherent efficiencies of smelting 中的 efficiencies，说的就是"高效率"。(3)前面提到了接近100%的金属得以回收，所以 nearly complete metal recovery 中的 complete 就可以直接译成"100%"，以回应前句。

第26段：(1)虽然原文有 and the reader is referred to...，但翻译时可以不翻译 the reader，中文省略主语的情况比较常见。后面的 for further details，当然可以译成"想获得进一步细节"，也可以译成正式一点的"如欲了解细节情况/详细信息"。

第27段：(1) consumable 意思是"可消费的、会用尽的"，这里译成"容易消耗的"，因为冶炼 1 吨电解铝要消耗 400kg—500kg 的碳阳极。(2) 注意 that forms the lining of the cell，既然这里的 forms 是第三人称单数，那么 that 只能指 a carbon cathode。

第28段：(1) In this process 就是"使用克罗尔法时"。(2) that contains molten magnesium 中的 contains 不是"含有"的意思，应该译成"装有"。

第29段：(1) 第二句 by adding zinc dust, ... in which case zinc is dissolved into the solution and gold is precipitated, or by adsorption onto activated carbon 是并列关系，不可误将 by adsorption onto activated carbon 译成 in which case... 定语从句的一部分。(2) 这段中的 gold，译成"黄金"也可以，不过考虑到这篇文章的文本类型，译成"金"也许更好。

第30段：如前所述，生产铜和镍时有时会有 slime，里面还有金、银等。as well as from copper and nickel smelting via the anode lime 指的就是"(从)铜和镍熔炼时产生的电极淤泥(中生产出来)"。

第30、31段之间的小标题：Secondary Sources 按字面意思译成"二次来源"意思不清，不妨译成"再生原料"。

第31段：(1) 第二句中 objects 和 items 都比较概括，需要译成涵盖面较大的词语，如"物件""物品"。(2) The recycled batteries are first separated... 这里的 separated 应被译为"分拆"，而不是"分离"或"分解"，因为这就是我们处理旧电池时发出的动作。(3) primary smelters 和 secondary smelters 中的 smelters 有"冶炼厂""冶炼炉"的意思，结合具体语境，此处应该理解为"冶炼炉"。(4) the sulfide concentrate charge 中的 charge 应该理解为"装料"，也就是"装进一次冶炼炉的硫化物精矿"。(4) sweeps 意思是"打扫"，这里引申为"打扫到一起的尘屑"，译为"尘屑"。

第32段：(1) the costs of producing concentrate and thus of primary metal production will continue to increase 可以译成一句话"精矿生产以及因之的原生矿石生产金属的成本将不断上升"，但有头重脚轻的感觉，不妨分成两句话来译。(2) 单纯从语法结构来看，最后一句中 during the primary production 可以理解为在修饰 the recycling of used metals and the improvement of product yield，也可以理解为只修饰 the improvement of product yield，考虑到 primary production 主要适用于原生矿石，所以理解为只修饰 the improvement of product yield 也许更准确。

第33段：(1) 第一句中的 individual metals 虽然字面意思是"单独的""个别的"，但从前文看，作者讨论了铜、镍、钛、铝、金、银等不少金属的生产，所以这里可以译成"部分金属"。(2) 翻译这一段时，需要增加一些连接词，以便保证译文连贯，如在第一、二句之间增加"但是"，在第三、四句之间增加"因此"等。

第34段：(1) 第一句有两个 increasing，翻译时应避免重复用同一个词语。(2) 最后一句的 applications，只译成"应用"的话，无法与 semiconductors、ultrapure metals 等词语搭配，译成"应用领域"更好。翻译表达时，一定要注意搭配问题。

第 35 段：既然后面有 a chemical method or a physical method，前面的 refining and purification methods 中的 methods 就无须翻译了。

第 41 段：(1)第二句中的 technique 和第三句中的 process 指都是"区域精炼技术"，为了译文的连贯，应译成同样的"技术"。(2)第三句中的 is held 不好译，尽管意思清楚，但表达费劲。肯定是"持"，但如何持？考虑"持"的是"棒"或"杆"，还要求相当稳，不能滑动，所以译成"夹持"。

第 44 段：(1)The method of fixing this gas 可以按照字面意思直译成"固定这种气体的方法是将其转化成硫酸"，也可以意译为"解决这个问题的方法是将二氧化硫气体转化为硫酸"。(2)the acid plant 译成"酸工厂"，读来不顺，考虑到捕获的是二氧化硫，用来生产的是"硫酸"，这里不妨译成"硫酸工厂"。(3)A further consequence 中的 consequence 一般译成"后果""结果"，但"后果"有贬义，"结果"不太贴切，可以意译成"好处"。(4)Being a corrosive and hazardous material 中的 hazardous 可以简单译为"有害的"，也可以译为"对环境和人体有害"。

第 45 段：(1)...contain up to 1% copper, often less 需要先译 often less，然后再译 up to，同时为了表达顺畅，需要在 up to 的译文前加"但也可能"几个字。(2)In the latter case 可以译为"在后一种情况下"，也可以直接译为"闪速吹炼的情况下"。

第 46 段：(1)hot gases 译成"热气体"也行，但准确地说这是一种"蒸汽"，所以最好译成"热蒸汽"。(2)waste heat boilers 可以译为"废热锅炉"，也可以译为"余热锅炉"，两者是一个意思。(3)metal values 字面意思是"金属值"，这里就是"一些金属"。

第 47 段：最后一句的 contain，既可以直译成"容纳"，也可以意译成"处理"。

第 48 段：(1)第一句中的 involves 不能译为"涉及"，可以译为"与……密不可分"。(2)dirty 译成"肮脏的"也行，但总觉得"肮脏的"带点主观色彩，所以译为"脏乱的"。(3)the process vessels 需要译成"生产中用到的容器"。在英语中，名词可以修饰名词，表达起来非常简洁，但这样的结构译成汉语时，必须添加修饰的词语，比如(4)最后的 transfer 不好译，可以译成"流动"，但"流动"只能描述液体、气体，而 materials 也可以是固体；译成"运输"，暗含需要额外动力，但如果是液体，有时靠自身动力就可以了。这里也许可以译成"传来送去"。

第 49 段：(1)从语境中分析，这段中 individual metals 应该不包括本文中讨论过的一些金属，所以这里可以译成"其他金属"。(2)are described in other appropriate articles in this encyclopedia 可以直译成"将在本百科全书的其他相应章节"。

参考文献：论文或著作中的参考文献本就是方便读者深入查找相关文献，翻译成中文反而给读者查找文献增添了困难，所以无须翻译，直接复制就可以了。

四、参考译文

有色金属：生产和历史

孙洪龙[①]

1 有色金属指铁以外的所有金属，以化合物(矿物)，如硫化物、氧化物、硅酸盐或碳酸盐的形式存在于地球的地壳。最重要的有色金属都以硫化物或氧化物的形式存在，常伴有铁，但有些金属，如金，也会以不含杂质的纯元素形式存在。虽然人们发现过大块的金子，但自然界中大多数金属和矿物都是以细小的粉粒散布在脉石矿物(如硅石或硅酸盐)中。在地球上的有些地区，这些金属或矿物的含量足够多，可以经济地进行开采。金属的富集是由各种地质运动造成的，金属和矿物的主要来源参见文献(如 Hayes, 1993; Dennis, 1965)。

2 最早为人们所使用的金属是铜，铜从矿石冶炼最初开始于公元前4000年的中东。最早使用的铜矿石很可能是相对较纯的氧化铜矿物，如接近地表发现的孔雀石，冶炼可能使用的是木炭床。

3 这些容易发现的矿石很快就用尽了，(因此)人们开始使用从更深矿床挖掘出来的包含铜的硫化物和其他金属尤其是砷的矿物。含砷铜矿物的冶炼生产出最初的青铜，带来了(人类历史上的)青铜器时代。后来，有毒的砷被锡取代，生产出了人们更熟悉的锡青铜，这种青铜的铸造性能、延展性能、可加工性能及硬度等综合性能都要好很多，因此成为接下来几千年里最常使用的金属。青铜器时代，金和银也被加工成装饰品，单独使用或给青铜制作的物件镀上金或银。

4 整个青铜器时代以及在接下来的铁器时代，人们也使用其他金属。将硫化方铅矿放置在木炭床上加热得到铅，铅被罗马人广泛用于管道工程和(制作)白镴餐具，(结果)无疑造成了很多铅中毒。通过使铅氧化，将银从铅中分离出来(灰吹法)，铅成为银金属的来源之一。从古希腊时，人们便从硫化矿朱砂中提取汞，因为汞可以同金形成一种汞合金，所以人们将汞用来给青铜镀金。公元前1世纪，在罗马，人们给铜加上28%的锌制造黄铜硬币，这一点的实现可能是通过将铜与碳化锌矿物一起加热实现的，因为锌金属自身是直到中世纪后期才被分离出来的。

5 中世纪黑暗时代，冶金领域没有任何值得称道的进展。随着文艺复兴的到来，三本重要的冶金书籍被编著出来：《火法技艺》(*De La Pirotechnia*)(Biringuccio, 1540)、《金

[①] 美国犹他州盐湖城犹他大学教授。

属学》(*De Re Metallica*)(Agricola，1556)和《矿石和矿石鉴定导论》(*Berschreibung Allerfürnemisten Mineralischen Ertzt und Bergswecksarten*)(Ercker，1574)，其中《金属学》一直是18世纪以前的标准参考书。这三本书籍描述了早期的矿石鉴定、冶炼、提纯、合金化及其他技术。

6　到18世纪，使用焦炭代替煤作为还原剂生产金属的工艺被成功开发。18世纪后期，人们发现了氢，很快便将其使用在金属氧化物(例如氧化钨)的还原上，生产纯金属(Habashi，1969)。

7　归功于电的发现、电池及后来电加热炉的成功研制，人们在19世纪发现和分离了很多别的金属。这个时期，电力在有色金属生产中的重要使用包括铜的电解精炼、霍尔-埃鲁特电解槽的发明用于生产铝。到20世纪中期，大多数现代有色金属生产技术的基础已牢固打下。想了解这个话题的最新概况及有色金属生产的基本原理，请参照阅读作者本人的专著(Sohn，2014a)。

从初始原料生产有色金属

8　从矿石中生产金属离不开一系列的物理和化学分离过程。通常，矿石(首先)需要破碎、研磨，以便从脉石成分中解放矿物颗粒。然后，利用矿物和脉石在比重、磁性或表面性质的不同采用几种物理方法中的一种将矿物和脉石分离开来。最常用的金属富集方法是泡沫浮选法，该法基于矿物和脉石固体表面在接触空气和水时的不同的亲和性，将气体(通常是空气)泡泡从包含细小固体颗粒的泥浆底部吹入，一些固体颗粒便附在气泡上，随气泡升到泥浆上部，而将更多的亲水颗粒留在泥浆底部。富集的精矿然后经化学处理，获得纯金属。化学处理可能是在高温下进行的(火法冶金)，也可能是在低温液体(通常是水)溶液中进行。(金属)提取方法的确定取决于很多因素：

- 矿物中各种金属的理化性质；
- 意图得到的金属的理化性质；
- 矿物中的金属含量；
- 矿物中的杂质含量；
- 相关的脉石矿的性质；
- 意图生产的金属的价值，以及
- 副产品的价值。

9　硫化矿非常适合用浮选的方式富集起来，得到的精矿通常采用火法冶金处理，个别硫化矿除外，相关讨论见下文。氧化物不易采用浮选的方式富集，因此被直接浸出或采用重力分选或磁选等其他方式富集起来。

10　硫化矿矿石除杂质外常含丰富的贵金属。火法冶金可以回收这些贵金属，因为贵金属不易氧化，与目标金属一起被生产出来，但贵金属性质不活泼，给湿法冶金处理硫化矿回收贵金属造成了障碍，像铋、钴、钼、硒、碲、铟、镉、锗等杂质也会被大量回收(Sohn，2014b)。

11 计划生产的金属的性质很大程度影响其提取方法。熔点相对较低(约小于1550℃)的金属适合火法冶金熔炼，熔炼时，脉石成分和容易氧化的杂质可能会迅速地移进熔渣，熔渣形成一种单独的液相，与含金属的不相容，可以快速移除。但是，生产高熔点的金属，首先要在中等温度下得到或气态、或液态的中间化合物，然后通过适当方式将中间化合物提纯，再还原生产出通常呈海绵状的金属。钛、锆和钽都属于这类金属。

12 低沸点的金属常以气态形式被回收，比如锌，用碳还原锌氧化物时，温度在锌的沸点之上，所以得到的是气态的锌。如果气态产品含有杂质，可能会通过分解蒸馏的方式分离出来(Sohn and Olivas-Martinez，2014b)。

13 从矿石提取金属时，金属的化学反应性是另外一个重要因素。化学反应性相对较低的金属如铜、镍、铅、锡可以在相对温和的环境下提取。如果这些金属出现在硫化物矿石中，提取金属是通过在熔融相中将矿石与氧气发生反应实现的。如果是氧化物矿石，则可以通过矿石与碳或氢的反应迅速实现。这些金属也可以通过将矿物浸出到水溶液中然后提纯、浓缩最后再通过电解沉积(常用)或沉淀(偶尔)从溶液中回收。

14 化学性质较活泼的金属如铬、锰在自然界中以氧化物形式存在，可以用碳还原，但不用氢。化学性质最活泼的金属像钛、锆、铝，其氧化物不能采取碳还原的方式，因为这样会生成碳化物，而不是金属。这些金属的生产工艺讨论见下文。

15 矿石中金属的含量影响着金属提取方法的选择。火法冶金工艺要求足够高品位的精矿，比如铜精矿只有含铜量超过约25%才适合熔炼。较低品位无法经济地富集到这个品位的矿石，需要先收集在一起成一堆，然后采用原地浸矿法通过将浸液长时间循环流过矿堆而回收。只出现在低品位矿石中的金属如金、铀，通常都是用浸出的方式生产出来的。

16 最重要的有色金属(铜、镍、锌、铅)的主要矿源都是硫化物。用火法冶金工艺从硫化矿中生产金属需要去除矿石中常存在的硫和铁，去除硫和铁靠氧气的氧化，硫的氧化形成二氧化硫，铁的氧化形成氧化铁(Sohn，2014a，b)。二氧化硫常被转换成硫酸，氧化铁则被吸收进废渣而被倒掉。在这个过程中，凝聚相处于熔融状态，因此该过程被称为"熔炼"。铜和镍的熔炼过程分两个阶段。第一个阶段即造锍熔炼，大部分硫和铁被氧化，留下混杂的金属硫化物和部分铁，这种混杂硫化物被称为"锍"。造锍熔炼时，大多数杂质进入矿渣，锍与矿渣分离，再进一步氧化，去除剩余的铁和硫，这种进一步处理被称为"吹炼"。吹炼出来的产物是一种相对较纯的金属，但仍然含有一些硫、氧和其他一些微量成分，需要在高温下精炼(火法精炼)最后再电解精炼提纯。电解精炼时，粗金属被用作阳极，提纯后的金属沉淀在阴极，杂质掉落到电解池的底部，通过进一步处理，回收有价值的资源，主要是贵金属。

17 有色金属产业使用的造锍熔炼和吹炼工艺有很多种(Sohn，2014b；Sohn and

Ramachandran，1998)最常使用的造锍熔炼工艺为闪速熔炼。闪速熔炼时，硫化物精矿与一种矿渣成分(通常是二氧化硅)同富氧空气一起被喷吹进闪速炉，发生氧化反应。因为氧化反应产生大量的热，所以一旦点火，反应就无须外界热量自发进行下去。处于熔融状态的锍和渣滴滑落进炉底(沉淀池)、凝结，然后分离成不会混合的层。

18　总的来说，造锍熔炼工艺可分为两类：悬浮熔炼和熔池熔炼。闪速熔炼就是悬浮熔炼的一个例子。熔池熔炼时，硫和铁的氧化过程发生在熔池中。另外一种类型的悬浮熔炼被称为漩涡熔炼，携带有精矿和熔剂颗粒的气体流被喷进漩涡反应器，气体与颗粒的交互作用更加迅速，反应过程较闪速熔炼更激烈，但要付出的高昂代价是反应器材料的腐蚀和其他操作上的难题。

19　熔池熔炼时，精矿和熔剂或者从熔池顶部装入，或者同反应气体一起经喷射器喷入熔体。不同工艺使用不同类型的喷射器。顶插浸没式喷枪将炉料喷吹到熔池液面以下，顶吹射流喷枪将炉料喷吹到熔池液面上方，风口插入式喷枪将炉料喷吹进熔锍的侧面或渣层，气罩式底吹喷枪将反应气体和固体颗粒垂直从炉底喷入。某些情况下，造锍熔炼在电炉里实现。

20　造锍熔炼过程中产生的熔锍从炉渣中分离出来，经吹炼成为粗金属。最重要的熔锍吹炼工艺是皮尔斯-史密斯吹炼。皮尔斯-史密斯吹炼时，使用钢包将熔锍从造锍炉运输至皮尔斯-史密斯转炉。皮尔斯-史密斯转炉为卧式，呈圆筒状，顶部有一个开口，侧面有一系列风口，空气经风口被吹入熔锍，引起氧化反应。虽然皮尔斯-史密斯工艺自19世纪后期以来一直是业界的标准，但在如今对工业生产要求日益严格的背景下，这种工艺在大气污染、人力需求及生产效率上都存在缺点。熔锍的运输和装料、熔融金属产品的倒出都会释放二氧化硫和其他有害气体。因为熔锍必须立刻处理，所以皮尔斯-史密斯转炉的运行必须与造锍熔炼炉的运行很好地协调，否则一个设备运行出问题，另一设备就得停炉等待。

21　克服这些困难的一个发明是肯尼科特-奥托昆普闪速吹炼工艺(Sohn，2014b；Sohn and Ramachandran，1998)。该工艺将造锍熔炼中生产出来的熔锍制成颗粒，研磨成细颗粒，然后堆放。这些金属硫颗粒在类似闪速熔炼炉但更小的闪速吹炼炉里吹炼。这种吹炼工艺的主要优点是吹炼连续且封闭，与造锍熔炼脱钩，能够产生一个连续的富含二氧化硫的气体流，方便生产硫酸。使用该工艺，二氧化硫的排放被大为降低，总体硫的捕获超过99.9%，相比之下，采用其他工艺，总体硫捕获率在80%至99%。

22　其他吹炼工艺是在配备有顶插浸没式喷枪或顶吹射流喷枪里实现的，过程与上面讨论的造锍熔炼类似。

23　锌和铅的硫化物通常先在一种被称为"焙烧"的环节中被氧化成氧化物，然后用碳作为还原剂将氧化物还原。焙烧得到的氧化物以固体形式被装进各种各样的反应器，反应器类型取决于促成还原反应发生的工艺(Sohn and Olivas-Martinez，2014a，b)。两步法之外,还有一种商业化的 QSL 铅生产工艺。使用该工艺,氧化和还原反应发生在一个长长的、水平的圆柱形容器的相对两端,氧化铅呈液态。氧化通过气罩底部喷吹器喷入的氧气实现,还原通过类似的底部喷吹器喷入氧气—煤混合气产生高温还原气氛实现(Sohn and Olivas-Martinez, 2014a; Sohn and Ramachandran, 1998)。

24 湿法冶金工艺生产有色金属需要将精矿或矿石中的金属浸出至水溶液,溶液经提纯、浓缩然后被分离成包含各种元素的液体流,最后再经电解(电解沉积)或沉淀法从溶液中回收金属。生产铜金属时,湿法冶金工艺主要适用于氧化或部分氧化的矿石或低品位的矿堆。与相对简单、更为迅速的火法冶金工艺化学反应相比,湿法冶金生产时发生的化学反应更复杂,也更慢,需要大量的溶液。其他问题还包括回收贵金属困难——因为贵金属无法浸出,以及会产生包含硫化物的稀溶液,需要处理后才能倒掉。世界上生产的铜有将近15%采用的是湿法冶金工艺(Habashi, 1997)。生产镍时,需要更多的湿法冶金环节,尤其是在矿石含铜量多的时候,浸出剂通常是硫酸盐、氯化物或氨水介质(Habashi, 1997)。世界上大部分锌的生产靠的是湿法冶金处理,将锌的硫化矿焙烧成氧化矿,然后浸出至酸性介质,提纯,再通过电解沉积回收锌。

25 尽管存在不少有意义的湿法冶金技术用于有色金属生产,但企图开发低成本、高效益的湿法冶金工艺以取代大规模火法冶金工艺还是遇到了失败。熔炼天然的高效率使得很难找到一个更好的替代工艺。这种高效来自于这样的事实:熔炼时,粗精矿(本身)就被用作有价值的燃料,提供生产需要的热量,只需要使用空气中的氧气即可,熔炼过程在高温下快速发生,金属几乎100%被回收,而所有这一切只需要相对不多的步骤。接近100%的金属回收率包括贵金属和其他副产品,而这对于湿法冶金来说要难很多。火法冶金的缺点包括:熔融金属盛装起来有困难;环境严苛,要求有严密的监控和技术娴熟的操作人员;精炼厂金属的库存很大。

26 非硫化物矿石中提取的重要有色金属包括轻金属(例如铝、钛和镁)、贵金属(例如金、银及铂族金属)、难熔金属(例如钨、钼、铌和锆)、碱金属(例如锂、钠和钾)以及放射性金属(例如铀、钚和钍)。这些金属的提取方法跟它们的赋存情况和性质一样差异巨大,如欲了解细节情况,请参阅百科全书和其他书籍的相应章节(Sohn, 2014a; Habashi, 1997; Gill, 1980; Gilchrist, 1980; Ryan, 1968)。这里将只讨论本类别下部分金属的主要生产方法(Sohn, 2014a)。

27 同多数性质活泼的金属一样,铝在自然界中以某种氧化物形式存在。含有铝的矿物通常在化学性质上相当稳定,因此很难还原成纯金属形式。对于铝的生产来说唯一重要的矿石为铝土矿,这是一种水合铝氧化物的混合物($Al_2O_3 \cdot H_2O$ 和 $Al_2O_3 \cdot 3H_2O$),通常含有氧化铁和二氧化硅等杂质。采用拜耳法从铝土矿中生产出氧化铝,其过程包括:使用苛性碱溶液溶出铝氧化物得到 $NaAlO_2$,然后继续水解得到 $Al(OH)_3$,使其沉淀;焙烧 $Al(OH)_3$ 得到氧化铝 Al_2O_3,再将氧化铝电解还原,生成金属铝。电解是在霍尔-埃鲁特电解槽中完成的，电解槽内有熔化的冰晶石(Na_3AlF_6)、AlF_3和CaF_2的混合物，氧化铝在槽中溶解。电极包括一个容易消耗的碳阳极和一个碳阴极，后者形成了电解槽的内衬。还原后的铝金属在电解槽的底部积存。

28 另外一种重要的轻(活性)金属是钛。含钛矿石主要有金红石和钛铁矿。自然界中发现的金红石主要含TiO_2(49.5%)，此外还有少量的氧化铁、二氧化硅和其他金属氧化物。钛铁矿主要成分是$FeTiO_3$，含35%~60%的TiO_2，可以通过电熔炼或其他化学方式处理增加其TiO_2含量。这些钛矿最主要的用途(大约95%)是生产TiO_2颜料。要生产钛金属，含有TiO_2的原材料需要在流化床内有焦炭存在的情况下氯化，首先生成$TiCl_4$，经蒸馏提纯。从$TiCl_4$中生产金属钛最常用到的方法是克罗尔法。使用克罗尔法时，$TiCl_4$被缓慢加入至一个装有熔融镁的反应器，生成海绵钛和$MgCl_2$。$MgCl_2$和多余的镁经蒸馏脱去，海绵钛经真空电弧熔炼固结。克罗尔法也被用于从用锆英石($ZrSiO_4$)中生产锆。

29 贵金属的生产值得一提。金的生产主要通过将矿粉用氰化钠溶解以便金以$Na[Au(CN)_2]$形式进入溶液。溶液过滤后，通过添加锌粉，使锌溶解进溶液，金析出得到金属金，或者通过吸附到活性碳上回收金属金。活性碳吸附时，负载的碳用苛性氰化物溶液洗脱，再次溶解金，然后通过电解沉积回收。金也作为铜冶炼的副产品被大量生产。铜精矿中的金跟铜一起经过整个熔炼过程，在铜电解提纯过程中与铜分离，与其他微量元素一起以淤泥的形式掉落至电解槽底部。

30 取决于含银矿物的化学特性，银的生产有几种不同的工艺(Habashi, 1997)。前面描述的用于金的生产的氰化法是从真银矿生产银的主要方法。大量的银也从铅和铅锌矿以及铜和镍熔炼时产生的阳极淤泥中生产出来。

从再生原料生产有色金属

31 跟铁和钢一样，有色金属也可以从再生原料中生产出来。主要的再生原料是废金属，来自初次生产或物件的制造以及回收的物品。铝工业在利用回收来的金属(大多数是旧饮料罐)生产初级铝方面一直非常成功。再生铅主要来自废旧铅酸电池的循环利用，约占除中国外全球铅总产量(2010年约为400万吨)的75%。这些回收来的材料可以由使用铅鼓风炉或QSL工艺①的一次冶炼炉加工，也可以由使用回转炉或反射炉的二次冶炼炉加工处理(Sohn and Olivas-Martinez, 2014a)。回收来的电池先被分拆成金属(Pb)和含有硫酸铅和氧化物的膏状物。在一次冶炼炉，膏状物与硫化物精矿混合。二次冶炼炉用碳将膏状物还原成金属铅。金大量地被从用过的含有其他金属如铜、镍、银等的金合金中回收，这

---

① QSL工艺或QSL法(Queneau-Schuhmann-Lurgi)是德国鲁奇(Lurgi Chemi)公司于20世纪70年代研究开发的直接炼铅工艺，技术基础是Queneau和Schuhmann两位教授的专利。QSl法是根据两位教授和鲁奇公司名字的第一个字母命名的。QSL工艺的关键设备OSL炉为可90°转动的卧式长圆筒形炉，并向放铅口方向倾斜0.5%。

些金合金来自于贵金属加工工厂生产的尘屑，或者来自大多由电子工厂产生的含金的涂层（Habashi，1997）。二次银的生产同二次金的生产原材料来源类似，此外还要加上回收的照相材料。

32　随着富矿的日渐减少，原生矿石的品位在下降，生产精矿成本将不断上升，从原生矿石中生产金属的成本也将不断上升。此外，考虑到要为越来越严格的环境保护付出的不断增加的成本以及不断增加的能源成本，旧金属的循环利用和初级生产中产品收率将变得越来越重要。

价值增值和精炼过程

33　到现在为止，我们讨论了有色金属的初级生产，也就是最广泛应用需要的最高纯度的部分金属的生产。但是有些应用要求金属有更高的纯度，有些应用需要不同比例的金属混合物，例如，电子工业使用的铜在某些情况下就要求纯度超过99.999%，氧含量需要低于0.001%，而相比之下，典型的电解提纯工艺得到的铜的纯度在99.97%至99.99%之间，含氧量为0.01%—0.03%。（因此，）为了满足某些应用的需要，通常需要（对金属进行）合金化处理，以便得到合适的性能，如耐磨性、耐腐蚀性、硬度、韧性、高温强度、颜色及电学和磁学性能。

34　随着科技不断进步带来的对更新、更先进材料需求的持续增长，通过改善现有材料的性能以及无疑通过发现具有新性能的新材料提高增加值已经成了有色金属工业生产的重要方面。这样的例子可以在半导体、超纯金属、化学气相沉积金属薄膜、金属间化合物和复合材料等应用领域找到。

35　为了生产性能严格符合要求的金属材料，需要得到成分波动范围极小和杂质含量极少的混合物。很明显，这就要求初始材料必须足够纯。金属精炼和提纯可以通过化学的或物理的方法来实现。

36　通过形成化合物从而去除杂质利用了这样一个事实，那就是不同的金属对于氧、硫和其他非金属元素有不同的亲和力。这样的化合物可能是固体，也可能是液体或气体，带有金属不同的物理性质，这些性质可用在金属的分离上。

37　化合物的形成通常在熔融相中实现，这是因为如下因素：
- 在液相中金属和复合物形成剂的接触可以立刻实现；
- 精炼后，可以方便地添加合金形成元素，并均匀混合；
- 熔融产品可以容易地成为各种终端形状。

38　通过形成化合物精炼和提纯金属的一个很好的例子是高纯铅的生产，其过程包括将初级冶炼中生产出来的粗铅进行几次化学处理（e. g., Sohn and Olivas-Martinez, 2014a; Hayes, 1993; Habashi, 1997）。

39　有些化学精炼方法先从要精炼的金属中形成一种挥发性化合物，将其与粗金属分开,然后再将挥发性化合物分解得到纯金属。分解环节通常会得到气态反应物,又可回收用于最初的挥发性化合物形成环节。这样的例子如用于镍提纯的蒙德(羰基)法和精炼锆、钛和钍等金属的碘化物法。

40 真空精炼利用的是不同元素、不同化合物的不同挥发性。虽然真空精炼主要用于钢的精炼以降低氢和氧的含量到低水平,真空精炼在有色金属的提纯上也有用途。

41 区域精炼用于从通常纯度已经相对较高的金属中制备纯度非常高的金属。该技术利用的是杂质在固相和液相时不同的溶解度(Pitt, 1979; Hayes, 1993; Habashi, 1986)。使用该技术时,在惰性环境中夹持一根初始材料制成的棒或杆,使其保持垂直或水平状态,棒或杆的一段采用某种方式(如感应加热器或电子束)形成一个狭窄的熔融相区域(Yu et al., 1992; Habashi, 1986),沿着棒长或杆长的方向移动棒或杆,杂质通常在液相中溶解度更大,因此集中在移动的熔融区域,(随着棒或杆的移动),被带至棒或杆的顶端。通常为了生产足够高纯度的金属,需要让金属在一个方向上多次经过熔融区域。这种技术主要用于半导体材料的生产,几乎任何金属都可以使用这种技术精炼。

42 电弧熔炼被用来精炼和铸造高熔点的活性金属,如钛、钽、锆和钼。在电弧产生的高温下,杂质通过挥发被去除。因为显而易见的原因,电弧熔炼必须在高真空或惰性气氛中进行。电极可能是由惰性材料或要精炼的金属制成。如果电极是由要精炼的金属制成,则电极会熔化,经过精炼,在水冷坩埚里形成金属锭,水冷坩埚起着另一个电极的作用。电渣精炼可以被归入此类别,除了渣池放置在精炼金属锭之上以影响精炼过程外。

43 沸点相对较低的金属如锌、汞,如果组成成分沸点差异较大,可以通过简单的蒸馏提纯,如果组成成分沸点差异不大,则通过分馏提纯。分馏基本上就是在一个分馏塔内根据需要重复蒸发、冷凝多次的过程,分馏时,蒸汽和液相向相反运动。如此,沸点较低的金属就集中在汽相中,反之亦然。

废弃物和环境问题

44 同其他工业一样,有色金属的生产会带来潜在的污染物。硫化矿生产有色金属时,会产生大量的二氧化硫,如前所述。解决这个问题的方式是将二氧化硫气体转化为硫酸。在世界上很多地方,二氧化硫排放进大气层仍然是个严肃的问题。但是,得益于更严格(环保)条例的执行,这种局面正在迅速改变,结果带来了更先进冶炼技术的研发和应用。事实上,有色金属行业研发和采用的大多数新技术,主要的驱动因素都是对环保的考量。最显著的改善是硫捕集量的迅速增加。产生逃逸二氧化硫气体或低浓度二氧化硫气流的老技术不断被新技术替代,这些新技术能够以匀速产生较高浓度的二氧化硫气流(Sohn, 2014a, b; Sohn and Ramachandran, 1998)。这使得可以在硫酸生产工厂捕获较多的硫。技术发展的另一个好处是有了更好的条件固定硫元素,而不是硫酸。硫酸的用途很多,是一种重要的化学产品,但硫酸腐蚀性强,对环境和人体有害,不容易储存和运输。如果在冶炼炉里大量生产，难以找到一个有吸引力的市场。二氧化硫以单质硫的形式被捕获，有很多好处，因为硫在室温下是固体且呈惰性，可以很容易储存、处理或在需要时随时随地转化为包括硫酸在内的有用产品(Kim and Sohn, 1999；Kim, 1999)。

45　金属的富集过程会产生包含矿石中脉石矿的尾矿。尾矿要么被运回矿山(采空区充填)，要么存放在尾矿池。冶炼过程产生的矿渣含有通常少于但也(可能)高达1%的铜。这些矿渣缓慢冷却后，经研磨和浮选处理，以渣精矿的形式回收其中的铜金属，然后循环利用于冶炼过程。在某些情况下，冶炼渣在电炉里进行处理，回收铜锍，剩余的矿渣作为废渣被倒掉。吹炼渣含铜可能高达10%，而采用肯尼科特—奥托昆普闪速吹炼工艺，矿渣中含铜量约为18%。闪速吹炼的情况下，矿渣经造粒、研磨、混入已经装在闪速熔炼炉的精矿料中。清理冶炼渣的成本可占铜冶炼总成本的15%(Sohn and Ramachandran, 1998)。冶炼过程中降低渣中铜含量、提高从渣中回收铜的效率，是重要的技术挑战。

46　来自熔炼和吹炼炉的热蒸汽含有相当多的热量值，在废热锅炉中得到回收。含尘气体通过废热锅炉被充分冷却后，通常用电除尘器将粉尘收集起来。粉尘中含有一些金属和杂质，再被送回熔炼炉。

47　湿法冶金工艺一直被视作火法冶金工艺的一种替代，尤其是从污染的角度来看。然而，(使用湿法冶金时)尽可能减少气体和烟尘排放的问题被处理含有酸和其他污染物的浸出渣和大量废水的问题所取代。湿法冶金生产带来的固体废物是矿渣，这些都是相对无害的，并且可以很容易地在有限的土地上处理。相比之下，容纳液态废弃物的难度要大得多。

48　一个常见的认知是，初级金属生产行业是与脏乱的工厂、简单的技术密不可分，(但)这一点远非正确，尤其是对于现代化工厂来说。(在现代化的工厂里)金属生产的很多环节都是由洁净环境下的现代仪器和计算机控制，生产中用到的容器已经很密闭了，各种材料在封闭的管道里不停地被传来送去。

49　其他金属的生产和各种金属提取技术的细节，请参阅本百科全书的其他相应章节。

五、专题讨论

科技翻译中的衔接和连贯

　　除学校的翻译练习和部分标语、口号、公示语的翻译外，大多数情况下，翻译不仅需要将原文中的词语译成目的语且组合成句，而且需要将译出来的句子连成文章。连句成文的时候需要考虑文章内部的衔接(cohesion)和连贯(coherence)。

衔接和连贯属于篇章语言学的范畴。根据篇章语言学，一个语篇（discourse）不是一堆杂乱无章的句子的叠加，而是段与段、句与句的有机组合，不仅是个有形的词语网络，而且是个无形的语义网络，服务于语篇的交际目的。前者即衔接，后者即连贯。

衔接是语篇特征的重要内容，体现在语篇的表层结构上，表现为语法手段（如照应、替代、省略）和词汇手段（如复现、同现关系和连接词）的使用。相比之下，连贯指语篇中语义的关联，存在于语篇的底层，通过逻辑推理来达到语义连接。①

一个连贯的语篇一般都具有衔接成分，句与句之间在概念上存在联系，句与句的排列符合逻辑。例如：

A common perception has existed that the primary metal production industry involves dirty plants and unsophisticated technologies. This is far from being true, especially for modern plants. Many of the steps in metal production are controlled by modern instruments and computers in a clean environment, and the process vessels have become much more enclosed with continuous transfer of materials through confined conduits.

虽然这个段落不长，但使用了不少衔接手段，如：This 指代的是前面的 A common perception；dirty 和 clean，modern 和 unsophisticated 属于比较照应；metal、production、plants、modern 等词语重复使用；所有的句子都使用现在时；等等。这些衔接手段构成了段落的有形网络。此外，组成这个段落的三句话在语义上关系十分紧密：第一句说的是大家的普遍认知；第二句是对这种普遍认知的评价；第三句是提供证据证明第二句的评价是正确的，即大家的普遍认知是错误的。

研究衔接和连贯对于译者（包括科技译者）来说非常必要。虽然汉、英两种语言都要求文章必须语义连贯，但在实现连贯的方式、方法上有区别。英语更多的是靠使用有形的衔接手段；汉语虽然也会使用衔接手段，但更多依靠的是语义之间的自然联系。所以总的来说，英译汉时通常要减少衔接手段的使用，而汉译英时则需要增加。

大多数熟悉汉、英两种语言的人都知道，英语重形合（hypotactic），汉语重意合（paratactic）。所谓形合，就是大量使用各种衔接手段；所谓意合，就是非必要不使用衔接手段，完全靠语义内在的联系连接。

衔接手段主要有五种。

第一种为照应（Reference），分人称照应（Personal Reference，表现为人称代词的使用，如 he、she、it、they、him、you、I、me、her、them 等）、指示照应（Demonstrative Reference，表现为指示词的使用，如 here、there、this、that、those、the）、比较照应

---

① 黄国文. 语篇分析概要[M]. 长沙：湖南教育出版社，1988：10-11.

(Comparative Reference，使用一些表比较的词语，如 better、more、different、else、same、identical、equal、similar、additional)等。

在照应的使用上，汉、英的主要区别有：

(1)英语中常用人称代词，汉语则尽量少用，一般通过重复名词来避免使用。

(2)英语中有关系代词，汉语中没有。

(3)英语中有 the，汉语中没有对应的词语。

(4)英语中常用 that 指代刚发生的事或刚说过的话，用 this 指较远的将来，汉语则分别用"这"和"那"。此外，汉语中的"这""那"还可指心理上的近或远。

这些区别，要求我们在翻译时注意转换表达手段，比如将英语人称代词、关系代词译成汉语名词，或省略不译。例如：

Oxides are not easily concentrated by flotation, and thus **they** are directly leached or concentrated by other means such as gravity or magnetic separation.

氧化物不易采用浮选的方式富集，因此被直接浸出或采用重力分选或磁选等其他方式富集起来。

原句中的 they 就不需要翻译。

Iron is one of the most common elements on earth. Nearly every construction of man contains at least a little iron. **It** is also one of the oldest metals and was first fashioned into useful and ornamental objects at least 3,500 years ago.

铁是地球上最常见的元素之一，几乎人类制造的每一个物件都或多或少包含铁。铁也是最古老的金属之一，至少在 3500 年前，人类已首次将铁加工成有用的物品或装饰品。

原段落第三句中的 It 最好不译成"它"，而译成"铁"，重复 It 指代的 Iron。

Sinter is made of lesser grade, finely divided iron ore, **which** is roasted with coke and lime to remove a large amount of the impurities in the ore.

烧结料是用品位较低的铁矿石破碎成细粉、同焦炭和石灰一起焙烧去除矿石中大量杂质而制得。

原句中的 which 可以不译。

第二种为替换。替换指用一个词替换另外一个词或句子成分，避免重复。英语中用来替换的词主要有三类：名词性的 one/ones、same 等；动词性的 do/did/have done 等；副词性的 so。翻译成汉语时需根据具体语境，灵活处理，one 和 ones 可以重复名词，same、so、do 等一般可以使用"同样的""也是这样""同样如此""同一"等表达方式。例如：

Many cell phones allow users to surf the Web, but only some newer ones are capable of wireless connection to the local area computer network.
很多手机都可以让用户在网上冲浪，但只有一些新型号的手机才能以无线方式连上计算机局域网。

Measurements of density and composition implied that Venus originally formed out of basically the same stuff as Earth.
密度和成分测试表明，金星最初诞生于基本上跟地球一样的物质。

第三种衔接手段为省略。汉、英双语中都有省略的使用，区别在于汉语更多地省略主语和宾语，而英语更多地省略谓语。例如：

Matter can be converted into energy, and energy into matter.
物质可以转换为能量，能量也可以转换为物质。

原句的后半部分省略了 can be converted，但翻译时需要翻译出来。

要获得纯锌是困难的。
It is difficult to obtain pure zinc.

汉语原句省略了主语，英语中用 It 作形式主语，to obtain pure zinc 不定式是真正的主语。

第四种是重复。汉语和英语中都使用重复这一衔接手段，但汉语中使用得较多，英语句子中的代词很多时候需要译成汉语中重复的名词，前面已有例子，此处不再赘述。

第五种，也是最重要的篇章衔接手段，即连接词的使用。英语和汉语都如此，但相对而言，汉语中连接词用得少一些，必要时才用。无论是英译汉还是汉译英，都可能需要添加连接词。比如本节开头的例子：

A common perception has existed that the primary metal production industry involves dirty plants and unsophisticated technologies. This is far from being true, especially for modern plants. Many of the steps in metal production are controlled by modern instruments and computers in a clean environment, and the process vessels have become much more enclosed with continuous transfer of materials through confined conduits.

一个常见的认知是，初级金属生产行业是与脏乱的工厂、简单的技术密不可分的，但这一点远非正确，尤其是对于现代化工厂来说。在现代化的工厂里，金属生产的很多环节是由洁净环境下的现代仪器和计算机控制的，生产中用到的容器已经很密闭了，各种材料在封闭的管道里不停地被传来送去。

译文中的"但"就是连接词，原文中并没有对应的词语，但暗含这个意思。

These easily found ores soon ran out, and minerals from deeper deposits containing sulfides of copper and other metals, especially arsenic, began to be used. Smelting of these arsenic-containing copper minerals produced the first bronze, thus bringing in the Bronze Age. Eventually, the toxic arsenic was replaced by tin to produce the more familiar copper-tin bronze, which has a much better combination of castability, ductility, workability, and hardness, and thus became the standard metal of use for the next thousand years. During the Bronze Age, gold and silver were also worked into ornaments, singly or to gild objects made of bronze.

这些容易发现的矿石很快就用尽了，因此人们开始使用从更深矿床挖掘出来的包含铜的硫化物和其他金属尤其是砷的矿物。含砷铜矿物的冶炼生产出最初的青铜，带来了人类历史上的青铜器时代。后来，有毒的砷被锡取代，生产出了人们更熟悉的锡青铜，这种青铜的铸造性能、延展性能、可加工性能及硬度等综合性能都要好很多，因此成为接下来几千年里最常使用的金属。青铜器时代，金和银也被加工成装饰品，单独使用或给青铜制作的物件镀上金或银。

译文中"因此"也是增添的连接词，增添的理据同样是原文中带有这个意思。
除了连接词之外，有时为了句子的连贯，还要增加一些其他词语，比如前面两个例子中的"在现代化的工厂里""中用到的""人类历史上的"①。
再来看一个汉译英的例子：

这些化合物还会在高炉上升管和下降管上凝结成结块，最终导致炉顶煤气通道受阻，或反应生成 HCl，腐蚀煤气净化系统的钢结构。
These compounds can also condense in uptakes and downcomers where they form

---

① 严格说来，不增加"人类历史上的"译句也说得通，但增加后意义更明了。

accretions that can eventually restrict the passage of the top gas, or react to from HCI and attack the gas cleaning system steelwork.

译句中的关系副词 where、关系代词 that 和最后一个 and 都起到了连接的作用，而原句中并没有与这些词对应的词语。

总之，无论是英译汉，还是汉译英，译者在表达阶段都需要关注译文是否通顺连贯，必要时需要通过增补衔接词语或调整句序的方式，使译文通达顺畅，符合逻辑。

◎ 练习十　将下面的句子翻译成汉语。

1. Most of the reactions used in pyrometallurgy are reversible, and so they will reach an equilibrium where the desired products are converting back into the reactants as quickly as the reactants are forming the products.

2. The equation for Gibbs free energy (ΔG), a measure of the thermodynamic driving force that makes a reaction occur, is:

$$\Delta G = \Delta H - T\Delta S$$

where ΔH is the enthalpy change in the reaction, T is absolute temperature, and ΔS is the entropy change in the reaction.

3. In the last century, low-temperature hydrometallurgical techniques have been developed which have been very complimentary to the conventional smelting methods, as they allow the treatment of the low-grade copper oxide ores that often occur along with copper sulfide ores, as shown in Figure 14.

4. Although aluminum is an extremely common element on earth, it was not practical to produce aluminum metal at a reasonable cost until two breakthroughs had been made: a method for producing purified aluminum oxide from bauxite (the Bayer process), and a method for converting aluminum oxide to metallic aluminum (the Hall-Heroult process).

5. Both rutile and ilmenite are hard, dense minerals that resist weathering, and as a result they tend to concentrate in placer sand deposits where they are mined by dredging, and separated from other valuable minerals by a combination of density, magnetic, and electrostatic separations.

6. Examination of the Ellingham diagram shown in Figure 23.2.2 illustrates why the metals known to the ancients were mainly those such as copper and lead, which can be obtained by smelting at the relatively low temperatures that were obtainable by the methods available at the time in which a charcoal fire supplied both the heat and the carbon.

7. Further chemical and mineralogical analyses, as well as mass balance calculations, are

currently under way, and aimed at determining whether any fluxes were used, more specifically whether iron or quartz were added to the charge to help form the fayalitic slag.

8. In fact, electrochemical mechanisms play a key role in each of the processes that are involved in obtaining the component metals from their sulfide ores, namely, prospecting for ores (weathering of sulfide ore-bodies), separation and concentration of valuable components of ores (flotation), dissolving the valuable metal content of ores (extractive metallurgy), recovering metals from solution (electrowinning), and purifying the metal (electrorefining).

9. Not only does the flotation process enable valuable sulfide minerals to be concentrated from worthless gangue minerals that make up the ore matrix, but it also allows different sulfide minerals to be separated from each other.

10. Environmental restrictions on sulfur gas emissions in smelting operations, and the limited market for sulfuric acid, resulted in a search for hydrometallurgical routes from sulfide minerals to metals.

附录一　汉英翻译练习

将下面的句子译成英语。

1. 1856年8月，一位名叫亨利·贝塞麦的英国人公布了他的炼钢方法，这个工艺最终能把钢的成本**降低到**原来的七分之一，更重要的是**使得大量生产钢成为可能**。

2. **之所以被称为**西门子—马丁平炉工艺，是因为铁水盛在一个比较浅的炉底或炉膛内，**如图1-1所示**。

3. 高炉炼铁的原料**可分为**以下几类：含铁原料、燃料及熔剂。

4. **按照化学成分**，铁矿石分为氧化物、硫化物、碳酸盐等，如表2-1所示。

5. 大部分铁矿石含铁量仅为50%—60%，因为它们还含有10%—20%主要**由**氧化铝和二氧化硅**组成**的脉石。

6. 与块矿和烧结矿**相比**，球团矿的优点是：粒度变化范围不大、质量稳定以及还原时具有良好的透气性。

7. 高炉喷煤可**大幅度降低**焦比，减少对日益匮乏的焦煤资源**的依赖**，是炼铁降低成本的最**有效手段**，已成为高炉炼铁技术进步的一项重要内容。

8. 原料从烧结厂和焦化厂被送入料槽系统，在这里，烧结矿、球团矿和焦炭被筛分和称重，然后**由**皮带或上料车进行上料。

9. 熔渣通常直接送到出铁场**附近的**渣池，在那里用水喷射冷却。

10. 混风阀在每一循环的开始打开，在热风炉送出的风温与预期的风温**相等**时关闭。

11. 如果不采取相应的措施，海绵铁在处理和储存过程中就**面临**重新氧化的危险。

12. 世界范围内，废钢**占**炼钢原料的40%，所以应该**被看作**一种重要的原料。

13. 就**如**铁水的生产**一样**，炼钢时使用造渣剂会产生一种反应性好、能吸收杂质元素的低黏度炉渣。

14. 平炉工艺曾经几乎**占**粗钢产量的100%，现今已缩减到可忽略的程度。

15. 氧吹炼钢工艺**主要是通过**使用高纯度的氧气，对以铁水（熔融生铁）和废钢组成的金属炉料进行精炼，以便快速生产含碳量和温度**合适的**钢水。

16. 目前氧气顶吹炼钢工艺（也称LD或LD/AC顶吹工艺）更被普遍采用，但在新钢厂中逐渐被复合吹炼工艺所代替。

17. 尽管一些硫会作为气体逸出，但是它一般会与生石灰**直接反应形成**硫化钙。

18. 氧气顶吹转炉呈梨形，形状与贝塞麦转炉和托马斯转炉**相似**。

19. 成渣剂和其他添加剂**以两种不同形式**加入：（1）这些材料**在吹炼**过程**中**由转炉顶部连续加入；（2）部分材料**在吹炼开始**加入，**剩余部分**在吹炼过程的几分钟内加入。

20. 氧气射流对熔池表面的**作用**应避免金属和渣的喷溅对炉衬造成的危害。

21. 在典型的氧吹工艺中，吹炼时间为10—20分钟，加上装料与出钢、出渣、测温以及取样时间，每炉次生产时间为30—50分钟。

22. 就工具钢、模具钢和不锈钢这些高级合金钢而言，必须**在严格控制的条件下**进行精炼，这样才能把杂质**减少到最低程度**。

23. 当渣中的氧化物与熔池中的碳反应时就产生了一氧化碳气体，**引起热沸腾**，**造成**氢气、氮气以及非金属化合物以气体形式逸出。

24. **温度较高时**，高炉中锰被还原的比例增加，但**在多数情况下**，还原的锰占装料进炉的锰的**比例在**65%—75%左右。

25. **结果是**，锰的分配比为炉缸热状态的**指标**，而**如下显示**，硅的分配比是更敏感的指标。

26. 对于特定的炉料和炉渣成分，铁水中硅含量和铁水温度**成正比**。

27. 磷的还原用下面反应式**表达**：

$$P_2O_5 + 5C \xrightarrow{\text{高温}} 2P + 5CO; \Delta H = +995.792 \text{ kJ/kmol}$$

28. 砷在高炉中的表现跟磷**很像**，即几乎完全还原并溶于铁水。

29. 钢液和炉渣系统之间的反应平衡在实验上和理论上**得到了广泛的研究**。理论研究包括热力学和物理化学研究。

30. 当高炉料线下降到一定水平时，下密封阀打开，炉料以一定流量流出，**进入布料溜槽**。

31. **计算表明**，提高喷煤比时，风口的横断面积需扩大，以便将风口风速维持在稳定操作所需的范围内。

32. **简单地说**，COREX工艺将炼焦炉和高炉工艺中执行的冶金步骤**按另一种顺序**完成，因此能使用非炼焦煤，但生产的铁水质量相同。

33. 随后炉顶煤气在洗涤塔中冷却、净化，然后作为高价值的输出煤气，可用于**很多**目的，包括发电、生产直接还原铁、用于高炉气体喷吹等。

34. 使用少量焦炭**以便**在停炉前和开炉后氧气缺少时稳定生产。

35. 为了控制由煤和其他原料**带来的**每吨11千克的硫含量，典型的炼钢厂要花费每吨钢超过5美元的费用，**另外还要**投资设备建设和特殊工艺，以控制钢产品中的硫化物。

36. 通过炼钢方式脱硫**将**钢中的硫含量**控制在合适的范围内**，非常昂贵，因此开发出了成本相对较低的铁水预处理技术，以便在氧气转炉炼钢前脱硫。

37. 该技术现在得到了广泛的应用，**能够**混合的粉剂范围很宽，且在生产过程中能够**允许**调节各成分的输送速度，如图 3-4 所示。

38. 另一个向铁水包中加脱硫剂的方法也**值得**注意。

39. 目前**有**多种不同的复合吹炼工艺，这些工艺在底吹气体种类、能达到的流速和吹入设备方面**不同**。

40. 现有的复合吹炼转炉是由传统的氧气转炉改进而来，其炼钢能力**在**60—300 吨**之间**。

41. 当电弧炉生产工厂**试图**进一步取代联合生产钢厂时，一些问题也**出现了**，如钢中残余元素含量和气体含量问题。

42. 据估计，采用高泡沫渣操作时，废气流量增加 1.5 **倍**，废气带走的热量增加 2.5 倍。

43. 图 3-17 **中的**钢包炉配备有碱性炉衬和水冷炉顶。

44. RH 工艺的原理如图 3-20 所示，而 RH 设备及其附属设备如图 3-21 所示。

45. Fujii 等人研究了流体流动对 RH 脱气装置脱碳速度的影响，并建立了其与各设计参数例如连通管和真空室直径及循环流量**的关系**。

46. Sewald **讨论了**RH 脱气装置的控制模型，**重点是**计算机监控系统及过程模型的设计和操作特点。

47. 这与 Palchetti 等人报道的数据**基本一致**，对于 300 吨钢包吹氧流量大约 70 Nm^3/min时，他们观察到了相近的再升温速度。

48. 炼钢添加剂**指的是除**主要合金组分**外**添加到钢水中的、用于形成或促进钢的良好性能的物质。

49. 图 8 至图 11 说明了以结晶器实际散热为依据的拉坯速度、钢水过热度和中间包浸入式水口偏流等最重要参数间的基本关系。

50. **解决**一些与异钢种多炉连浇相关的质量问题很重要。

51. 对钢样进行**分析**，然后加入铁合金**调整**钢水成分。

52. 由于铬对碳具有相当大的亲和力，故铬被归类为碳化物形成元素。

53. **在**烧结造块**的**过程中，挥发物由于受热而除去，石灰石熔剂和焦炭与铁矿石**结合在一起**，从而生产出**被**称为"烧结矿"的具有不规则形状的熔结团块富矿。

54. 鼓风的**主要**作用是**使**焦炭燃烧并产生很高的温度。

55. **由于**空气/氩气泡的抬升作用，结晶器中的钢水流动模式发生了变化。

56. 这台年生产能力90 万吨的板坯连铸机**装备有**自动调宽结晶器、大容量中间包、多点顶弯与矫直设备及气雾冷却系统。

57. **为使**板坯得到均匀冷却，现已采用了气/水喷雾冷却系统。

58. 钢的硬度、韧性和耐磨损及耐腐蚀性能可以通过添加合金元素、采用特种生产工艺以及热处理**得到改善**。

59. 碳是钢中重要的合金元素。**一般而言**，随着含碳量的增加，钢的强度提高，而韧性降低。

60. **从本质上说**，合金钢是通过添加合金元素改变其特性并**产生**预期力学性能变化的碳钢。

61. 伯利恒钢公司雀点厂的一台投产一年的板坯/大方坯兼用型连铸机于1987年2月创造了一项连浇炉数的世界**记录**。

62. 这座高炉炉缸直径44.25 ft，工作容积129914 ft^3，每昼夜能稳定生产9000美吨以上的铁水。

63. 为了预测作为断面尺寸和结晶器工作长度(金属弯月面至结晶器出口的距离)函数的临界拉坯速度，他们进行了进一步的**模拟试验**。

64. 常常采用退火热处理来**消除**冷轧加工硬化。

65. 在每一周期内，当结晶器向下运动的速度**超过**拉坯速度时即出现负滑脱。

66. 结晶器内钢水面上升到一定高度，以下两种操作随即**开始**：结晶器开始上下往复振动，引锭杆在连铸机内被向下拉出。

67. 冷却速度在一定程度上**决定**了铸铁是白口铸铁还是灰铸铁；冷却速度越快，形成白口铸铁的倾向就越大。

68. 一种**含有**0.15%的碳、0.40%的钼和0.003%的硼的钢，具有高强度和良好的焊接性能。

69. 如图9(b)**所示**，临界点含碳量约为0.688%(质量分数)，脱碳速度平均值约为0.102%(质量分数)/分，标准偏差分别为0.11%和0.01%。

70. 视冶金要求不同，VD法全循环时间为25—55 min。

71. 1996年，中国粗钢产量**超过**日本，成为世界上最大的产钢国。

72. 为了进一步提高商业和工业用污水槽、电站过热管线及大气污染区内建筑物保护层等所需的耐蚀性，还可添加**高达**2.5%的钼。

73. 金属经过抛光或刚切割后，会发出**特有的**"金属光泽"。

74. 该文献指出，氧气顶吹转炉使用脱硅铁水可使渣量减少、金属收得率提高。

75. **由于冷钢比热钢硬，因此冷轧钢机需要配备更硬的工作辊。**

76. **焊接型钢是具有开口横截面的条钢，它们的形状特点已由 6.2.1 条界定。**

77. 在设计结晶器窄面水缝时需要格外**注意**，以确保这些水缝足够延伸到窄面的边缘处。

78. 当钢水冲刷铸坯的凝固前沿时，铸坯壳就发生**部分重熔**。

79. 这些化合物在高炉的高温区挥发，然后在向炉顶上升的过程中凝结到冷却板周围，从而**引发**腐蚀。

80. 使用红外线测渣系统时，所有的炉前工，**不论**其技术水平高低或出钢时环境能见度状况优劣，都能精确地对出钢钢流裹渣进行测定。

81. 在吹氧期间，光学仪表系统能**连续**测量由炼钢转炉炉口发出的光线强度。

82. 钢水由钢包浇入到容量为 17 吨的中间包，**随后**通过浸入式水口（图 2）浇入带有足辊的长为 1 米的立式结晶器中。

83. 欧盟真空处理能力的迅猛增加，**反映**出人们对优质钢的需求。

84. 这些化合物还会在高炉上升管和下降管上凝结成结块，**最终**导致炉顶煤气通道受阻，或反应生成 HCl，腐蚀煤气净化系统的钢结构。

85. 为了保证两座高炉能够达到提高一代寿命和利用系数的主要目标，炉子耐火材料和冷却系统的设计都需要进行重大**改进**。

86. 这一点与较高的拉坯速度导致较少的保护渣消耗和较浅的震痕这一事实**相一致**。

87. 由于连铸在节能、金属收得率和生产率方面较之传统的模铸工艺具有极大的**优势**，因而连铸取代模铸是不可避免的。

88. 为了实现新钢厂生产的全连铸，需要进行**广泛的**开发研究工作。

89. 写作本文的目的是**预测**操作参数的改变对氧气顶吹转炉炉墙上的溅渣数量和位置的影响。

90. 一个细小的钢件，例如一个小型工具或一根针，可以在水中淬火，并且在整个断面上获得**均匀的**特性。

91. 电弧炉**开始熔化时**，电极被废钢包围，因而电弧对炉墙的辐射作用可以忽略不计。

92. 氧气炼钢工艺的主要目的是将熔池中的碳含量由 4% 左右降低到 1% 以下（通常为 0.1% 以下），降低或控制硫、磷含量，最终将钢水温度提高到 1635℃ **左右**。

93. 低于临界含碳量时，传质速度不足以维持与喷入的全部氧进行反应。

94. 一项耗资 8300 万美元的 1 号高炉大修和改造项目目前正**在进行中**。预计该项目完工后，日产量将提高到 400 吨，达到 4000 吨。

95. 这项投资的大部分将用于完成目前的**在建**项目，包括一条新的镀锌作业线、一台不锈钢冷轧机和一座高炉的大修。

96. 一般**认为**，钼的能力是钨的两倍，因此，过去含有 18% 钨的钢，现在可以用含有约 9% 的钼的钢来代替。

97. 这些不锈钢具有卓越的耐腐蚀和抗变形综合性能，在**高温**下仍能保持高强度。

98. 布料溜槽围绕炉子竖轴线**旋转**，并调整到预定的水平倾角。

99. 对中碳钢、低碳钢和超低碳钢连铸用保护渣进行了**取样**，以便比较其特性。

100. 电弧炉组件**分成**几个功能组：容纳废钢和钢水的炉体结构、用于炉体和主要结构件运动的组件、电弧炉供电组件及炉体上或周边的辅助设备。

101. 随着二次精炼技术的**出现**和大幅度提高钢的性能和质量的推动，中间包已变成了三次精炼器。

102. 在炉外精炼操作中最广泛使用的设备是钢包炉。钢包炉可以是简单改造的设备，也可以是**精心设计**的设备。

103. 炼铁工艺所需热量并非**完全**是由焦炭的燃烧提供，因为对于大多数高炉而言，大约有 40% 的热量是由热风的显热提供。

104. 使用两个独立供煤系统向高炉供煤：一个供煤系统向 20 个**偶数**风口供煤，另一个供煤系统向 20 个奇数风口供煤。

105. 连铸机总产量是过钢速率和浇注时间比的函数，同时它还和拉坯速度、流数、铸机生产计划、连浇比以及运行可靠性**有关**。

106. **如果要**对金属进行成形加工，例如进行深冲加工，最重要的是金属要有延性，**换言之**，金属应具有较高的延伸率。

107. 低延伸率金属在进行拉深加工时会产生断裂，因为在成形加工时所使用的金属要**承受**巨大的变形。

108. 现在有些钢厂采用的做法是每隔几周**定期**停产 8 小时更换炉壳。

109. **由于**采用了小辊距和短辊身，辊子晃动和磨损问题实际上已得到解决。

110. 由于铸铁具有良好的流动性和较低的凝固收缩率，故而可获得较高的精度公差和**相当大的**设计自由度。

111. 随着熔炼的**进行**，对废钢和熔池的传热效率下降，电弧辐射热更多地传向炉墙。

112. 2 号烧结机**流程示意图**见图 2-1，其主要特点列于表 2-1。

113. 一般而言，出钢末期炉渣检测**不是**由经验丰富的炉前操作人员用肉眼观测，**就是**借助安装在炉子上的电磁传感器进行。

114. 另一种避免炉渣卷入出钢钢流的实用方法是：当出钢钢流中发现有炉渣时即**终止**出钢过程。

115. 氩气搅拌时，渣/金属界面处的紊流会导致卷渣和钢水中外来夹杂物的**形成**。

116. 钢的连铸是在有色金属连铸法的基础上发展起来的，但在尝试采用有色金属连铸法进行连铸钢的初期却**受到**了一系列难题**的困扰**，其中最棘手的难题是凝固坯壳与结晶器壁的黏结。

117. 这一操作模式的另一优点是可以使用较长电弧，从而提高了电能效率，降低了电极单耗。

118. 由于直接**观察**和研究真实氧气顶吹转炉中的溅渣护炉过程极为困难，现已采用室温模拟的方法研究溅渣护炉参数。

119. 当装入高炉炉顶的炉料和焦炭**下降**通过高炉炉身时，它们被来自风口上升的炽热煤气预热。

120. 静态装料模型利用一个炉次的**起始**和终了数据即可计算出炉料装入量和氧气需

求量。

121. **此外**，浸入式水口对中不良、Al_2O_3 在其端部积聚都可能引起浸入式水口与钢质铸坯壳搭接，**从而使**坯壳破裂，**导致**漏钢。

122. 实践经验**表明**，提高枪位会降低转炉上部区域的溅渣量。

123. 吹氩处理后，运输小车将钢包运到合金添加/加热处理站，**在那里**对钢水进行合金化处理并加热。

124. 为减少水口堵塞，**广泛使用了**水口喷氩。

125. 目前正在**付诸实施**的带钢热轧机改造计划包括安装一套厚度和形状自动控制系统。

126. **尽管**出钢口平均**寿命**迅速提高，但其性能却不稳定。

127. 使用示踪剂测定、数模和水模方法**研究了**六流中间包内多孔缓冲板对钢水流动的影响。

128. **毋庸置疑**，薄坯壳会**增加**在振痕波谷处开始漏钢和渗钢的**几率**，特别是当还存在着一些诸如较高的过热度、较大的结晶器摩擦力或注流冲击等其他**不利因素**叠加在一起时更是如此。

129. 这就保证了将钢水池的搅拌面局限于钟形隔离罩下方的区域。

130. 将红热的管坯运至液压机处。液压机在管坯一端的中心压出一个锥形凹痕，**为下一步穿孔作业作好准备**。

131. 高炉中**发生**的化学过程既多又复杂，但是可以将它们归纳为最简单的形式：铁的氧化物+碳→铁+碳的氧化物。

132. 因此，无论过去和现在，在世界范围内都**进行了大量的研究**和开发工作，**旨在**探索替代高炉流程的经济和生态友好的方案。

133. 泡疤是指气体**未能**冲破熔态金属的表面张力而产生的一种表面缺陷。

134. 可以对钢表面渗硼，以便获得既坚硬又耐磨的表面，其效果常常**优于**渗碳或渗氮的表面。

135. 漏钢概率与结晶器出口处最低坯壳厚度**成反比**，与钢水温度成正比。

136. 锰本身是脆的，**作为工程材料**毫无用处，但是作为钢的合金元素，锰**有助于**抑制钢的脆性，提高钢的韧性。

137. 通过多孔耐火衬向中间包内喷入惰性气体是提高夹杂物去除能力的**潜在方法**。

138. 钢水熔池喷钙处理的**主要目的**是将固态 Al_2O_3 夹杂物转变为液态钙铝酸盐，以防 Al_2O_3 堵塞水口。

139. **人们发现**，高炉喷吹废塑料的**优点**包括既可降低焦比和二氧化碳排放，又可提高高炉利用系数和寿命。

140. 与氧气顶吹转炉炼钢法相比，氧气底吹转炉炼钢法的**一个独特优势**是它**能熔炼较**

大块废钢。

141. 如图 9(b) 所示，临界点含碳量约为 0.688 质量分数，脱碳速度**平均值**约为 0.102 质量分数/min，**标准差分别**为 0.11 和 0.01。

142. 负滑脱时间的**定义**为：在一个振动周期内，结晶器在拉坯方向上以大于拉坯速度运动的时间。

143. 由于中间包壳数量有限，因此中间包快速周转是维持这台连铸机持续运行的**关键**。

144. 平均浇铸速度为 33 吨/时/流，**额定**浇铸时间为 72 分钟/炉。

145. **依据**带钢厚度、铸辊尺寸和钢水池深入，拉坯速度典型的变化**范围**为 40～130m/min。

146. 机械加工性能是工程上考虑的一项重要因素。机械加工性能可以通过将钢中含硫量**提高**到 0.3% 得以提高，而通常情况下钢中含硫量大约为 0.04%。

147. 由于结晶器液面波动随拉坯速度的增加而加剧，因此，克服这一问题就成为实现较高拉坯速度的**前提**。

148. 采用离心铸造法生产的产品表面致密而均匀。因而，该法生产的排水管或气缸套质量**优于**砂型铸造的同类产品。

149. 陶瓷侧坝必须**保证**结晶器的密封性，**以防**钢水泄露和在陶瓷侧坝上凝固。

150. 木炭炼铁炉温度较低，故铁水吸收的硅量很低；同时由于木炭本身**不含硫**，故所炼生铁亦不含硫。

151. 在北美大多数炼钢厂，高炉铁水的含硫量为 0.040%—0.070%；而为了使钢水成分**符合**连铸机生产规定的成分范围和最终产品的质量要求，兑入氧气转炉的铁水含硫量仅为 0.001%—0.010%。

152. **此外**，由于粉煤喷吹量增加，炉料的矿焦比也增加，从而导致焦炭入炉数量相对下降。

153. 高速钢这种非凡的高温稳定性主要得益于钢中**存在**碳化钨和碳化铬。

154. **据说**，这种涂层经久耐用，因为基体金属与氧化物涂层的膨胀系数基本相同。

155. 软熔带的形状、高度以及焦窗厚度**取决**于装料实践、**矿焦比**（或焦比）和含铁炉料的软化和熔化温度。

156. 由于必须使用环境友好型焦炉取代旧式焦炉，因而研发下一代新焦炉有其**必要**。

157. 如果说需要一根具有良好外观和强度的钢管用于制造——**比如说**飞机结构件或皮下注射用针头，那么还需要进行冷拔加工。

158. 钢中的溶解氧与碳结合生成的一氧化碳气体在凝固前沿逸出。

159. **为了**进行比较，将类似尺寸的分段式出钢口袖砖的出钢时间和使用寿命示于图 6。

160. 这一点与较高的拉坯速度导致较少的保护渣消耗和较浅的振痕这一事实**相一致**。

161. 由于连铸在节能、金属收得率和生产率方面较之传统的模铸工艺具有极大的优势，因而连铸取代模铸**不可避免**。

162. **20 世纪的最后二十年见证了**连铸机设计的**巨大进步**。

163. 20 世纪 60 年代末，许多连铸坯的表面质量尚**不如**传统方法生产的铸坯。

164. 浸入式水口在保证板坯连铸操作、防止钢水注流吸氧吸氮和获得结晶器内理想钢水流动方面**起着至关重要的作用**。

165. 如果将**一大块钢**加热，然后淬火，那么，**尽管**这块钢外表冷却迅速，但其**内部**冷却十分缓慢。

166. 本研究将视角集中于那些能在结晶器金属弯月面区域**引起**冷却条件变化的因素，并**尝试解释**在一炉浇注期间菱形变形的程度与方向变化。

167. 对使用不锈钢进行的试验来说，根据数学模型计算出的结晶器内平均热**通量**与根据冷却水温度升高计算出的结晶器内平均热通量**十分吻合**。

168. 铌作为合金化组分主要用于耐热、抗蚀钢，对于**提高钢**的高温强度效果良好。

169. 防止漏钢的**对策**包括提高中间包容量、**改进**长水口设计、修改中间包位置误差、改进结晶器液位检测装置、选择合适的结晶器保护渣等。

170. 钢在油中淬火**比**在水中淬火**好**，因为水淬速度过快，会**导致**工件开裂或表层脱落。

171. 钢包钢水使用三根石墨电极加热，电极埋入渣层，由导流电极臂支撑。

172. 然而，在一个振动周期内振痕是如何形成的这一点上**尚未达成共识**。

173. **由表 6 给出的数据**可以看出，无论是过滤的钢水量还是非金属夹杂物去除效率都令人满意。

174. **本研究的最终目的**是增强对这些缺陷形成机理的认识，探索与铸坯壳自然收缩更加匹配的结晶器窄面的锥度设计，以期减少结晶器的应力与变形以及**其他相关问题**。

175. 钢水使用 Mn **部分脱氧**时，铁也**参与反应**，生产的脱氧产物为液态或固态 Mn(Fe)O。

176. 当装入高炉炉顶的炉料和焦炭**下降**通过高炉炉身时，它们被来自风口上升的炽热煤气预热。

177. 由钢包下渣到中间包对于正在浇注的钢水质量**危害极大**。

178. 布料溜槽围绕炉子**竖轴线**旋转，并调整到**预定的**水平倾角。

179. 脱碳期间，供碳分析用的**钢样取自**下降管的下面，取样**间隔**为 30s~60s。

180. 黏接漏钢时，钢水面**附近的一部分**坯壳**黏结**到结晶器壁上，随后与向下运动的坯壳**断开**。

附录二 汉英翻译练习参考答案[①]

1. In August 1856, an Englishman, Henry Bessemer, **made public** the description of his process of steelmaking which eventually **reduced** the price of steel **to** about a seventh of its former cost and more important still, **made it possible** to produce steel **in large quantities**.

2. The Siemens-Martin open-hearth furnace **was so called** because the molten metal lies in a comparatively shallow pool on the furnace bottom or hearth **as Fig. 1-1 shows**.

3. The raw materials for the production of iron in the blast furnace can **be grouped as follows**: iron-bearing materials, fuels and fluxes.

4. Iron ores **are classed by** their chemical compositions, such as oxides, sulfides, carbonates, etc., **as shown in** the table 2-1.

5. Most iron ores **contain** only 50 to 60 per cent iron because they contain 10 to 20 per cent gangue which **consists of** mostly of alumina and silica.

6. **Compared with** lump ores and sinter, the advantages of pellets are: a narrow size range, constant quality and good permeability during reduction.

7. BF pulverized coal injection can **dramatically reduce** coke rate and the **dependency on** increasing shortage of coke resource, so it is the most **effective approach to** reducing the ironmaking cost and has become an important part in BF ironmaking technology

8. Raw material from the sinter and coke plants are fed to a system of bunkers, where sinter, pellets and coke are screened and weighed, before being charged **via** a belt system or the skip car.

9. The molten slag usually goes directly to slag pits **adjacent to** the casthouse, where it is cooled with water sprays.

10. The mixer valve is open at the start of each cycle and closes progressively until the hot air leaving the hot blast stove **is equal in** temperature **to** the desired hot blast temperature.

11. Sponge iron **is exposed to** the hazard of reoxidation during handling and storage unless **corresponding measures are adopted**.

[①] 注意黑体字词语,这些词语在科技翻译中经常用到。

12. Worldwide, scrap **participates** 40% **in** steel production input and must therefore **be considered as** an important raw material.

13. Just **as in the case of** the hot metal, the slag formers are used in steelmaking to produce a reactionable low viscosity slag **capable of** absorbing undesired elements.

14. The open-hearth process, once **responsible for** almost 100% of raw steel production, **has now dwindled to** negligible proportions.

15. Oxygen steelmaking processes **are concerned mainly with** the refining of a metallic charge consisting of hot metal (molten pig iron) and scrap **through the use of** high-purity oxygen to rapidly produce steel of the **desired** carbon content and temperature.

16. The oxygen top-blown process (LD or LD/AC) is currently the more common, but is gradually giving way to combined blowing process in some new plants.

17. Sulphur will normally **react directly with** the burnt lime to **form** calcium sulphide, although some will escape as a gas.

18. Oxygen top-blowing converter is a pear-shaped converter **similar in** shape **to** the Bessemer and Thomas converters.

19. Slag formers and other additions are charged **by two different methods**: (1) These materials are given from the converter top continuously **in the course of** blow. (2) Part of these materials is given from the converter top **at the beginning of** blow and **the remaining quantity** is introduced during a few minutes in the course of blow.

20. The **effect of** oxygen jets **on** the bath surface should not **involve** splashing of metal and slag, which might be harmful for the lining.

21. In the oxygen process typically, blowing periods of 10 to 20 minutes **plus** filling and emptying plus temperature measuring and sampling together **result in** the tap-to-tap time of about 30 to 50 minutes.

22. For high-grade alloy steel such as cutting tools, die steels, and stainless steel, the metal must be refined **under rigidly controlled conditions** and **in such a way that** impurities are **reduced to a minimum**.

23. When the oxides included in the slag react with the carbon of the bath, this **gives rise to** the gaseous carbon monoxide which **causes** the heat **to** boil and hydrogen, nitrogen and non-metallic components escape as gases.

24. **At higher temperatures** the percentage of manganese that can be reduced in a blast furnace increases, but **in most practices** this **amounts to approximately** 65% – 75% of the manganese charged.

25. **As a result** manganese partitioning is **an indicator of** the thermal state of the hearth, and

as shown below, silicon partitioning is an even more sensitive indictor.

26. For any particular burden and slag composition the silicon content of the hot metal is **proportional to** the hot metal temperature.

27. The reduction of phosphorus **is expressed by** the reaction:

$$P_2O_5 + 5C \xrightarrow{\text{high temperature}} 2P + 5CO; \quad \Delta H = +995.792 \text{ kJ/kmol}$$

28. The behavior of arsenic in the blast furnace **is very much like** that of phosphorus, **in that** it is almost completely reduced and dissolved in the hot metal.

29. The reaction equilibria in the liquid steel-slag systems have **been extensively studied**, both experimentally and theoretically by **applying** the principles of thermodynamics and physical chemistry.

30. When the furnace stockline has descended to the desired level, the lower seal valve opens and **allows** the charge **to** flow at a controlled rate onto the distribution chute.

31. **Calculations indicted that** the tuyere cross-sectional area would have to be increased at higher PCI rates in order to keep the tuyere velocity within the established range needed for operating stability.

32. **In simplified terms** the COREX process **applies** metallurgical steps usually carried out in a coke oven plant and a blast furnace **in a different sequence**, thus **allowing the utilization of** non-coking coal, but producing the same quality of hot metal.

33. The top gas is subsequently cooled and cleaned in a scrubber and **is then available as** a highly valuable export gas **suitable for a wide range of** applications, including power generation, production of DRI, blast furnace gas injection, etc.

34. **A negligible amount of** coke is used to stabilize the process **in case of** oxygen shortages before and after shutdown.

35. To control the 11 kg/ton of sulfur **contributed by** the coal and other feedstocks, the typical steel plants spends over \$5/ton of steel **in addition to** capital charges for equipment and exclusive of processes for sulfide control in the steel product.

36. To **bring** the sulfur content of the steel **to within the range** manageable by the far more costly steel desulfurization, the lower cost hot metal treatment technologies have been developed to remove sulfur **prior to** the oxygen steelmaking step.

37. This technique, now in universal use, **allows for** a wide array of reagent combinations and **permits** independent adjustment of the rates of the delivery of the reagents during the process, Fig. 3-4.

38. Another method for delivery of desulfurization agents into hot metal transfer ladles is

worthy of note.

39. **There are** at present many different combination blowing processes, which **differ in** the type of bottom gas used, the flow rates of bottom gas that can be attained, and the equipment used to introduce the bottom gas into the furnace.

40. The existing combination blowing furnaces are converted conventional BOF furnaces and **range in** capacity **from** about 60 tons **to** more than 300 tons.

41. As the EAF producers **attempt to** further displace the integrated mills, several issues **come into play** such as residual levels in the steel and dissolved gases in the steel.

42. It is estimated that offgas flowrate can increase **by a factor** of 1.5 and offgas heatload by a factor of 2.5 for operation with high slag foaming rates.

→**factor**: If an amount increases **by a factor of** two, for example, or by a factor of eight, then it becomes two times bigger or eight times bigger. E.g.:

The cost of butter quadrupled and bread prices increased by a factor of five.
黄油的价格是原来的 4 倍,面包的价格上涨了 5 倍。

43. The ladle furnace **illustrated in** Fig. 3-17 is lined with a basic lining and covered with a water-cooled roof.

44. **A schematic illustration** of the principle of the RH process **is depicted in** Fig. 3-20 while **a sketch of** a RH unit with ancillary equipment **is shown in** Fig. 3-21.

45. Fujii et al. studied the effect of fluid flow on the decarburization rate in a RH degasser and **related** this **to** design parameters such as the snorkel and vessel diameters as well as the circulation flowrate.

→**et al.**: and others (used as an abbreviation of 'et alii' (masculine plural) or 'et aliae' (feminine plural) or 'et alia' (neutral plural) when referring to a number of people)

46. A control model for the RH degasser **was discussed by** Sewald **with emphasis on** the supervisory computer control system as well as the design and operational characteristics of the process model.

47. This **is in broad agreement with** data reported by Palchetti et al. who observed a similar reheating rate for an oxygen blowing rate of approximately 70 Nm^3/min for 300 tonne heats.

→**Nm^3**: 标立方(N = normal condition)

48. Addition agent for steelmaking **refers to** any material **other than** the principal alloying constituents, added to the molten metal to produce or promote **desirable** properties in the steel.

49. **Figs. 8 through** 11 show the fundamental relationships between the most important

variables of cast speed, steel superheat, and biased flow from the tundish SEN on actual mold heat removal.

→**SEN**：submerged entry nozzle（浸入式水口）

50. It is very important to **address** the quality implications **associated with** the sequence casting of unlike grades.

51. A sample of the steel **is analyzed** and **adjustments** to the composition **are made** by adding ferroalloys.

52. Since chromium **has a** fairly great **affinity for** carbon, it **is classed as** a carbide-forming element.

53. **In the process of** agglomeration, the volatile matter is driven off by the heat; the limestone flux and the coke **are incorporated with** the ore, and the agglomerate is produced **in the form of** irregular clinker lumps, **known as** "sinter".

54. The **primary function** of the air blast is to **enable** the coke to burn and produce a very high temperature.

55. The change in flow pattern of steel in mold **is due to** the "lifting effect" of the air/argon bubbles.

56. The continuous slab caster with **an annual capacity of** 900,000 tonnes **is equipped with** an automated variable width mold, large capacity tundish, multipoint bending and unbending devices and air mist cooling system.

57. An air/water mist spray cooling system has **been adopted** to **achieve** uniform cooling of strand.

58. The hardness, toughness and resistance to wear and corrosion of steels can **be modified** by adding alloy elements, special production processes and heat treatment.

59. Carbon is the significant alloying element in steel and, **in general**, strength increases but toughness decreases with increasing carbon content.

60. Alloy steels are **basically** carbon steels with additional alloy elements added to alter the characteristics and **bring about** a predictable change in the mechanical properties.

61. Bethlehem Steel's one-year-old continuous slab/bloom caster at its Sparrows Point Plant established a world sequence length **record** during the month of February 1987.

62. With a hearth diameter of 44.25 ft and a working volume of 129,914 cu ft, the blast furnace is capable of consistently producing **in excess of** 9000 NTHM/day.

63. Further **simulations** were performed by them to predict this critical casting speed as a function of section size and working mold length (distance from meniscus to mold exit).

64. In order to **eliminate** work-hardening after cold rolling, heat-treatment by annealing is

frequently applied.

65. A negative strip occurs when the downward velocity of the mold during each cycle **exceeds** the withdrawal speed of the strand.

66. At a certain level of steel in the mould two operations **are** immediately **initiated**: the mould unit begins to reciprocate in a vertical direction and the dummy bar is withdrawn through and down the continuous caster.

67. The rate of cooling to some extent **determines** whether the cast iron is white or grey; the more rapid the cooling, the greater the tendency to form a white iron.

68. One steel **containing** 0.15% per cent carbon, 0.40% per cent molybdenum and 0.003% per cent of boron has a high strength and good weldability.

69. **As seen in** part (b) of Figure 9, the carbon content at the critical point is about 0.688 weight percent and the mean value of the decarburization rate is about 0.102 weight percent/minute, with standard deviations of 0.11 and 0.01, respectively.

70. **Depending on** the metallurgical requirements, the complete cycle times for the VD process can range between 25 and 55 minutes.

71. In 1996, China **overtook** Japan to become the world's largest producer of crude steel.

72. Molybdenum, **up to** 2.5 per cent, is also added for still greater corrosion resistance required, for example, in commercial and industrial sinks, superheater tubing in power stations and the cladding of buildings in polluted atmospheres.

73. When polished or freshly cut, metals show **distinct** "metallic luster."

74. The literature **indicates** that slag volume decreases and yield improves when desiliconized hot metal is used in BOF.

75. **As** cold steel is harder than hot steel, cold steel mills need harder work rolls.

76. Welded sections are long products of open cross-section which have a shape characteristics **defined** in 6.2.1.

77. **Care** is especially needed in designing the water slots on the narrow face of mold to ensure that they extend close enough to the narrow face edge.

78. When liquid steel washes the solidification front of strand, the strand shell is **partially** remelted.

79. In the high-temperature zone of the blast furnace these compounds are volatilized and as they rise toward the top of the furnace they condense around cooling plates and **cause** corrosion.

80. With the infra-red slag detection system all furnace operators are able to detect slag in steel stream at tap in a precise way, **independent of** their skill level or visibility conditions during the tap.

81. The light meter system **continuously** measures the intensity of light emitted from the mouth of the steelmaking vessel during the oxygen blow.

82. Liquid steel is fed from the ladle into a 17-tonne capacity tundish and, **subsequently**, via a submerged entry nozzle (Fig. 2), into a 1-metre long vertical mold equipped with foot rollers.

83. The strong increase in vacuum treatment capacities in the EU **reflects** the demand for high-quality steels.

84. These compounds can also condense in uptakes and downcomers where they form accretions that can **eventually** restrict the passage of the top gas, or react to from HCI and attack the gas cleaning system steelwork.

85. To insure both blast furnaces would be able to achieve the primary goals of extended campaign life and increased productivity, design of the furnace refractory and cooling systems requires a substantial **upgrade**.

86. This **is consistent with** the fact that higher casting speeds give less flux consumption and shallower oscillation marks.

87. It is inevitable that ingot casting is replaced by continuous casting as the latter has tremendous **advantages** in terms of energy saving, metal yield and productivity over the traditional ingot casting process.

88. Going for 100% continuous casting of the new steel plant production requires **extensive** development work.

89. The purpose of this paper is to **predict** the effects of changing operation parameters on the quantity and location of slag splashed onto the wall of a BOF.

90. A thin piece of steel such as a small tool or needle can be quenched in water and will give **uniform** properties throughout its section.

91. **At the start of** meltdown in electric arc furnace the radiation from the arc to the sidewalls is negligible because the electrodes are surrounded by the scrap.

92. The main purpose of the oxygen steelmaking process is to reduce the carbon in bath from about 4% to less than 1% (usually less than 0.1%), to reduce or control the sulfur and phosphorus, and finally, to raise the temperature of the liquid steel to **approximately** 1,635℃.

93. Below a **critical** carbon content the rate of mass transfer is insufficient to react with all the injected oxygen.

94. An $83 million reline and upgrade of No. 1 blast furnace is **in progress** that is expected to increase productivity by 400 toones/day to 4,000 tonnes/day.

95. Much of this investment will be used to complete projects that are currently **under**

construction, including a new galvanizing line, a cold-rolling mill for stainless steel, and the relining of a blast furnace.

96. It is generally **reckoned** that molybdenum has twice the "power" of tungsten, so that a steel previously containing 18 per cent tungsten could be replaced by one with about 8 per cent molybdenum.

97. The stainless steels have an excellent combination of resistance to corrosion and deformation, and maintenance of high strength at **elevated temperatures**.

98. The distribution chute **rotates** around the vertical axis of the furnace and changes to predetermined angles with respect to the horizontal plane.

99. Fluxes used for the casting of medium carbon, low carbon and super ultra low carbon steel grades were **sampled** to compare properties.

100. The EAF components **fall into** several functional groups: furnace structures for containment of the scrap and molten steel, components which allow for movement of the furnace and its main structural pieces, components that support the supply of electrical power to the EAF, and auxiliary process equipment which may reside on the furnace or around its periphery.

101. With the **advent** of secondary refining techniques and the drive toward steels of significantly improved properties and quality, the tundish has become a tertiary refiner.

102. Ladle furnaces are among the most widely used pieces of equipment in secondary steelmaking operations and range from relatively simple retrofitted installations to **elaborately designed** facilities.

103. The heat for the ironmaking process is not produced **entirely** by the combustion of coke, because at most blast furnaces roughly 40% is supplied from the sensible heat of the hot blast air.

104. The coal is supplied to the blast furnace by two separate coal-supply systems: one feeding the 20 **even-numbered** tuyeres and the other feeding the 20 odd-numbered tuyeres.

105. Total caster output is a function of throughput rate and casting time ratio, which in turn **relates to** casting speed, number of strands, machine scheduling, sequence ratio and reliability.

106. If a metal is intended to be shaped, for example by a deep drawing operation, it is essential that it should be ductile; **in other words**, its elongation figure should be high.

107. A metal with a low elongation would crack when subjected to deep drawing, because the metal used **is subjected to** considerable deformation during shaping.

108. Some plants now follow a practice where the shell is changed out **on a regular basis** every few weeks during an eight hour downshift.

109. **Thanks to** the narrow roller spacing and the short roller barrel, the problems of roller wobble and wear have practically been solved.

110. The great fluidity of cast iron and its low shrinkage on solidification make possible close tolerances and **considerable** freedom in design.

111. As melting **proceeds** the efficiency of heat transfer to the scrap and bath drops off and more heat is radiated from the arc to the sidewalls.

112. A schematic flow diagram of No. 2 Sinter Plant is shown in Fig. 2-1 and some important features of this Plant are summarized in Table 2-1.

113. Generally, slag detection at the end of tap is carried out **either** visually by a highly skilled operator, **or** by the use of electromagnetic sensors installed on the furnace.

114. In the other practice avoiding furnace slag entrainment into the tap stream, the tapping process is **terminated** when slag is found in the tap stream.

115. Turbulence at the slag/metal interface during argon stirring can result in slag entrainment and the **formation** of exogenous inclusions in the steel.

116. The continuous casting of steel was developed on the basis of the continuous casting process for non-ferrous metals, but early attempts to cast steel by this process **were beset with** a number of difficulties, the most troublesome of which was the sticking of the solidified shell to the walls of the mould.

117. Another advantage of this mode of operation is that a relatively long arc can be **employed**, resulting in increased energy efficiency and lower specific electrode consumption.

118. Because it is very difficult to **observe** and study directly the slag splashing process in the actual BOF, room temperature modeling has been used to study splashing parameters.

119. When the burden materials and coke that are charged into the top of the blast furnace **descend** through the stack, they are preheated by the hot gases ascending from the tuyeres.

120. The static charge model uses **initial** and final information about the heat to calculate the amount of charge and the amount of oxygen required.

121. **In addition**, poor SEN alignment and alumina buildup on the tip may cause the SEN to contact the steel shell, and this **in turn** rupture the shell and **result in** a breakout.

122. Practical experience **indicates** that raising the lance height decreased the amount of slag splashed to the upper areas of the vessel.

123. **Following** argon-bubbling treatment, the transfer car moves the ladle to the alloy addition/heating station **where** liquid steel is alloyed and heated.

124. Argon-injection into the nozzle **is widely employed** to reduce nozzle clogging.

125. A hot strip mill improvement program **in progress** includes the installation of an

automatic gage and shape control system.

126. **In spite of** the rapid increase in the average taphole **service life**, its performance is not stable.

127. **An investigation** of the effect of multiple-hole baffles on the flow of steel in 6-strand tundish **was made** using tracer measurements, mathematical and water modeling.

128. **Undoubtedly**, thin shell can **increase the probability** of a breakout or bleeder to initiate at the base of an oscillation mark, especially when other **adverse factors**, such as higher superheat, higher mold friction or steam impingement are superimposed.

129. This **ensures** that the agitated surface of the steel bath **is confined to** the area underneath the bell.

130. The hot billet is passed to a hydraulic press which makes a cone-shaped indentation in the center of the billet's end, **making it ready for** the piercing operation **which will follow**.

131. The chemical processes **taking place** in the blast furnace are many and complex, but **bringing** them **to the simplest form**: oxide of iron + carbon→iron + oxide of carbon.

132. Therefore, **intensive research and development** worldwide **carried out** in the past and present **aimed at** an economical and ecological **alternative** to the blast furnace route.

133. Blister refers to a surface defect produced by gas that **fails to** break the surface tension of a metal when it is in the molten state.

134. Steels may be surface "boronized" to **give** a hard-wearing surface often **superior to** that produced by carburizing and nitriding.

135. The breakout probability **is inversely proportional to** the minimum shell thickness at the mold exit and **is directly proportional to** the temperature of liquid steel.

136. **On its own** manganese is brittle and useless **for engineering purposes** but, when alloyed in steels, it **helps to** combat brittleness and it increases toughness.

137. The bubbling of inert gas into the tundish via a porous refractory lining **offers a potential means to** enhance inclusion removal.

138. The **primary purpose** of a calcium injection into the steel bath is to convert solid Al_2O_3 inclusions to liquid calcium aluminates to **prevent** Al_2O_3 **from** clogging casting nozzles.

139. **It has been found** that the **benefits** from injecting waste plastics into blast furnace include reducing coke rate and emission of carbon dioxide, and increasing productivity and campaign of blast furnace.

140. **A distinct advantage** of the OBM process is its **capability to** melt bigger and thicker pieces of scrap than the BOF process.

→OBM: oxygen bottom-blowing method

141. **As seen in part** (b) **of Figure** 9, the carbon content **at the critical point** is about 0.688 weight percent and the **mean value** of the decarburization rate is about 0.102 weight percent/minute, with **standard deviations** of 0.11 and 0.01, **respectively**.

142. The negative strip time **is defined as** the amount of time during the oscillation cycle when the mold travels in the cast direction at a greater velocity than the cast speed.

143. Fast tundish turnaround **is critical to** keeping this caster operating due to the limited number of tundish shells.

144. The average casting rate is 33/tones/hr/strand with a **normal** casting time of 72 min/heat.

145. Casting speed **ranges** typically **from** 40 **to** 130 m/min **depending on** strip thickness, casting roll size and pool height.

146. Machinability, an important engineering consideration, can be **enhanced** by **increasing** the sulphur content of the steel to as much as 0.3 percent, compared with about 0.04 percent which is normal.

147. Since mold level fluctuations worsen with increasing casting speed, overcoming the issue **is a prerequisite** for attaining higher casting speed.

148. The centrifugal casting process **yields** a product having a dense, uniform outer surface; consequently, a drain pipe or cylinder liner cast by this method is considered **superior to** similar ones cast in sand moulds.

149. Ceramic side dams must **guarantee** the tightness of the mold **against** leakage of liquid steel and solidification on these dams.

150. The lower temperature of the charcoal ironmaking furnace results in a much lower absorption of silicon by the iron, and since charcoal **is free from** sulphur, the pig iron produced is also sulphur-free.

151. In most North American steel plants, the hot metal leaves the blast furnaces containing 0.040%–0.070% S, while the oxygen converters are charged with hot metal containing as little as 0.010%–0.001% S, to **conform to** limits on steel composition set by caster operations and final product quality requirements.

152. **What's more**, **owing to** the increase in the injection amount of pulverized coal, the ore/coke ratio of furnace burden has increased, **resulting in** a relative decrease in the charged amount of coke.

153. The **remarkable** stability of the high-speed steel **at high temperatures** is largely due to the **presence** of tungsten carbide and chromium carbide.

154. This coating **is said to** have high durability owing to the fact that both the underlying

metal and the oxide coating have similar coefficients of expansion.

155. The cohesive zone shape, elevation and coke slit thickness **are determined by** the charging practice, **the ore/coke ratio** (or the coke rate), and the softening and melting temperatures of the ferrous materials.

156. **There is a need to** develop next generation coking ovens because the old ovens will have to be replaced but only by environmentally friendly ones.

157. If a tube of optimum surface appearance and strength is required for, **say**, aircraft structures or hypodermic needles, a further process of cold-drawing is applied.

158. The dissolved oxygen in steel **combines with** carbon to give carbon monoxide gas, which is released at the solidification front.

159. **For comparison**, the tap time and service life of segmented taphole sleeves of similar **dimensions** are shown in Figure 6.

160. This **is consistent with** the fact that higher casting speeds give less flux consumption and shallower oscillation marks.

161. **It is inevitable that** ingot casting is replaced by continuous casting which has tremendous **advantages in terms of** energy savings, metal yield and productivity **over** the traditional ingot casting processes.

162. The last two decades of the twentieth century **saw a dramatic advance in** the design of continuous casting machines.

163. At the end of the 1960s, the surface quality of many continuously cast semis **was inferior to** that of conventional products.

164. Submerged entry nozzles **play a vital role in** ensuring continuous slab casting operations, preventing oxygen and nitrogen pick-up of steel stream and achieving a desired flow in the mold.

165. If **a massive block of** steel is heated and quenched **the interior of** such a block will cool fairly slowly, **despite** the rapidity of cooling on the outside.

166. This study **focuses on** events that can **induce** variable cooling conditions in the meniscus region of the mold and **attempts to explain** the variability in the severity and orientation of rhomboidity during a heat.

167. **The agreement between** the average heat fluxes in the mold calculated by the mathematical model **and** by the cooling water temperature rise **is particularly good** for the trials with the stainless steel.

168. Columbium is used as an alloying constituent principally in the heat and corrosion resistant steels where it **promotes** good high temperature strength.

169. The **countermeasures to** prevent breakouts include increasing the tundish capacity, **modification** of the shroud design, correction of tundish position error, improving mold level detecting device, selecting proper mold powder, etc.

170. The steel is quenched in oil, which **is much preferable to** water, since too rapid quenching may **lead to** cracking or peeling of the case.

171. The steel in the ladle is heated **with the aid of** three graphite electrodes which are in slag layer and supported by current conducting arms.

172. However, **there is not a consensus on** how a mark forms during one cycle of oscillations.

173. **From the data in Table 6, it can be seen that** both the amount of steel melt filtered and the removal efficiency of non-metallic inclusions **are satisfactory**.

174. **The ultimate goal of the research project is to** increase the understanding of the mechanism of formation of these defects, and to find a taper design for the narrow face of the mold that more closely **matches** the natural shrinkage of the shell, in order to reduce the stress and distortion of mold and other **related problems**.

175. When the steel is **partially** deoxidized with Mn, the iron also **participates in** the reaction, forming liquid or solid Mn(Fe)O as the deoxidation product.

176. When the burden materials and coke that are charged into the top of the blast furnace **descend** through the stack, they are preheated by the hot gases **ascending** from the tuyere.

177. Slag carryover into the tundish from the ladle **is extremely detrimental to** the quality of the steel being cast.

178. The distribution chute rotates around the **vertical axis** of the furnace and changes to **predetermined** angles **with respect to** the horizontal plane.

179. During the decarburization, **a sample** of the steel **was taken** for carbon analysis under the down-leg snorkel **at an interval of** 30-60 seconds.

180. In a sticker breakout, **a portion of** the strand shell **in the vicinity of** the liquid steel level **adheres to** the mold wall and **separates from** the downward-moving shell.

附录三　课后练习参考答案

练习一（略）

练习二

1. 钢铁工业是世界上最重要的工业之一，也是传统史上最古老的工业之一。早在公元3000年前，铁就是人类文化与文明的基础。从矿石中提炼铁的历史可以追溯到史前时期。古时候，铁矿石是在炭火中加热的（起初无疑是偶然的），当火熄灭时，像海绵一样的固体铁块就产生了。

2. 钢铁工业是世界上最重要的工业之一，也是传统史上最古老的工业之一。早在公元3000年前，铁就是人类文化与文明的基础。

3. 人们最早从陨石中获取铁，而陨石又是从天而降，所以"Iron"（铁）这个词源于古伊特鲁里亚语"aiser"意为"神灵"。

4. 当铁与适量（微量）的碳结合就生成一种比铁更强韧的金属：钢，人们用钢可以制成各种各样的物品包括餐具、战舰、摩天大楼甚至航天火箭。

5. 作为建筑材料，铁的最大缺点是容易与湿润的空气反应（此过程被称为锈蚀）并生成片状红棕色的人们叫做铁锈的氧化物。

6. 物质可根据其物理形态（物态）分类，例如可以分为气态、固态或液态。铁属于固态物质。铁也是一种过渡金属，这类金属共性为有延展性，可塑性，可导电和热。

7. 贝塞麦炼钢法是最早的，用比较经济的方式把铁水炼成钢的工业流程。此法通过向铁水中喷吹空气，降低碳含量生产低碳钢。

8. 从矿石中提炼铁的历史可以追溯到史前时期。古时候，铁矿石是在炭火中加热的（起初无疑是偶然的），当火熄灭时，像海绵一样的固体铁块就产生了。海绵铁能够锤打成型制造工具和武器。冶金前辈发现，在吹火或扇风时，火会燃烧得更旺，铁就会更快地炼成。从此以后人们就使用风箱来增加风量。

9. 工业革命以前，钢是一种贵重材料，只能少量生产用来制造像剑和弹簧这样的物件，而结构部件则用铸铁或熟铁来制造。

10. 战争极大地刺激了世界各国的钢铁产量。爆发在20世纪的两场世界大战对钢铁的生产产生了深远的影响，当时由于武器设备需求激增，钢铁厂及其他一些重工产业都被收归国有。

练习三

1. 天然磁石是指有磁性的磁铁矿,能吸引铁或钢。最具有经济价值的铁矿包括磁铁矿和赤铁矿。

2. 严格来说,铁是铁合金的一种,和铸铁比,它的碳含量低得多,和熟铁比,它们的碳含量相当(或略高)。在钢中添加其他金属元素可以为钢材增加其他性能。

3. 根据内蒙古某细粒低品位磁铁矿的矿石性质,以降低生产成本为出发点,采用阶段磨矿、阶段磁选工艺,可获得高品位铁精矿。

4. 对于形状比较复杂,而所用材料的可延性和可锻性又比较差,就需要经过一系列过程,才能将原料成形。

5. 可以认为,腐蚀是金属在接触湿气、空气和水等介质时所受到的破坏性化学侵蚀,铁生锈只是其中的一个例子。

6. 铁好像人一样也能呼吸空气,而且它一边呼吸空气,一边由铁矿石般坚硬变得松软,成为肥沃富饶的有益的尘土,进而成为我们赖以获取食物的土壤,成为我们建造房屋的石头,成为构成山脉骨架的岩石,成为环绕大海的沙滩。

7. 众所周知,在夜间蝙蝠通过回声定位来导航,但这种方法只对较短的行程有效,它们如何在长途飞行中定位,人们还不甚了解。通过研究我们发现人为改变地球磁场会影响大棕蝠的归家路线,从而证明这些蝙蝠是参照磁罗盘定位找到归家的路。

8. 板块漂移学是研究地球地壳及地幔结构的学科,研究前提是认为地壳分裂为大块的硬块(地球板块)并漂浮于半流体的岩石之上,板块之间于边界处相互碰撞挤压。其相关联理论有大陆漂移理论和海底扩张假说。

9. 岩浆中可以看出一系列的氧化条件。矿物对的组成可以用来计算岩浆是如何氧化的,还可结合分级结晶确定岩浆可能的演化过程。

10. 另外,通过研究磁铁矿和其他氧化物矿之间的相互作用,可以确定地质史上岩浆的当时的氧化环境和演化过程。

练习四

1. 波兰焦炭生产的动态发展要求采用有效的方法处理焦炭废水。

2. 一般情况下,当对材料施加载荷时,在高温下焦炭可能发生延伸、蠕变变形,其应力不断增加直至断裂。

3. 高浓度的污染物及其毒性导致在排放到水、地面和外部实体的污水系统之前需要进行多级处理。

4. 焦炭在高炉中发生反应,其强度降低。也就是说,焦炭暴露于高温的二氧化碳时,由于化学反应而发生降解。在工业评价中,反应后焦炭强度(CSR)也被用作强度指标。

5. 通过对多孔结构的研究，可以得出结论：所选粉尘吸附剂可作为焦化废水预处理的有效吸附剂。

6. 研究范围包括对选定的吸附剂（煤尘、焦炭、生物炭）的吸附性能进行检查，对生物处理后的焦化废水进行物理化学测试，以及旨在确定吸附剂剂量和建立过程平衡状态所需时间的检查。

7. 为测定与二氧化碳反应后焦炭的强度（CSR），取直径为 20±1 mm 的焦炭 200g，在 1100℃的温度下与 CO_2 反应 2 小时。CO 的流速为 5 NL/min。1100℃保温 2 小时。

8. 吸附技术在水和废水处理中发挥着重要的作用，因为它具有很高的效率并且在去除各种有害有机化合物过程中没有选择性。

9. 在炼焦过程中，配煤中高硫煤的比例较高，不仅会增加合成焦炭中的硫含量，导致高炉操作的改变和铁不合格，而且还会增加脱硫过程的负荷。

10. 在本研究中，为了研究在室温和高温下焦炭强度的差异，我们测量了高温 CO_2 反应气氛下的焦炭强度。

11. 煤的性质、矿物和硫的形式以及热解条件（温度、气氛和传质）是影响煤热解过程中硫转化行为的主要因素。

12. 由于 CO_2 反应降低了焦炭的重量，所以试样的重量会降低。不幸的是，断裂和化学反应产生的细颗粒的重量目前无法测量。

练习五

1. 经检测 COVID-19 患者存在肠道微生态失衡，表现为肠道的乳酸杆菌、双歧杆菌等有益菌明显减少。

2. 在钢铁生产过程中，高炉炼铁消耗的能源最多；因此，进一步降低还原剂的用量是非常重要的。

3. 值得注意的是，只有当杨氏模量或剪切模量对计算结果影响不大时，才能降低杨氏模量或剪切模量以提高计算效率。

4. 由于烧结矿和焦炭的形状不规则，给接触参数的测量和校准带来了困难。

5. 只有这样，才能获得更准确的不同炉料结构下的炉料流动、分离和分布数据。

6. 徐文轩等学者利用 DEM 建立了 4070 立方米的高炉模型，分别分析了烧结矿、球团矿、块矿在料斗内径向、周向以及垂直方向上的体积百分比。

7. 他们发现，当填充料斗时，小颗粒的炉料倾向于分布在中心，较少移动到料斗边缘。大颗粒的炉料与小颗粒的分布趋势相反；中间颗粒未发生偏析。

8. 通过风口表面的隔热来实现热损失的减少，通过热风混合燃料后混合体均匀性的改善来实现燃料燃烧充分性的提升。

9. 在高炉风口表面制造隔热层的已知方法包括在机头部分外表面等离子喷涂陶瓷层。但由于风口材料与喷涂层的持续粘附性不强，导致涂层脱落，因此该方法并未得到广泛应用。

10. 尽管研究人员已经对炉料的流动和分离进行了详细的研究；由于计算能力、模型算法、接触参数校准等方面的限制，现有的炉料流动与分离研究还有待进一步完善。

11. 为了提高天然气与热风的混合效率，建议在离空气通道轴线更近的地方供给天然气，并在几个地方同时使用多种类型的旋流器。考虑到空气通道内温度同时升高，要求喷嘴内部有良好的隔热性能。

12. 当炉料在旋转溜槽内运动时，小颗粒炉料主要分布在溜槽底部，大颗粒炉料分布在小颗粒上方，大颗粒的角速度在旋转溜槽末端较大。料轨外侧大颗粒较多，内侧小颗粒较多；大颗粒的角速度更大，更容易滚动。

练习六

1. 高炉中富含钛的铁矿石(如钛铁砂)会造成问题，因为碳化物和氮化物在炉渣中会形成固体颗粒。

2. 直接还原工艺使用气体或煤将铁矿石还原为固态海绵铁，而不是从高炉中生产高温熔融金属。

3. 铁矿石可以全部是球团，也可以是球团和块矿混合而成，通过锥形料斗送至熔炉顶部。

4. 第一步，将矿料加热，使用来自另一个反应器的气体还原剂进行预还原。还原将在该反应器中进行。

5. 冷却气体从冷却区底部引入，从顶部出去，在顶部经由洗涤器进行冷却和清洁。

6. 占主导地位的替代技术以直接还原和熔炼法为基础。直接还原和熔炼法在20世纪60年代末和70年代初出现，通常在低于铁熔点的温度条件下，不使用焦炭而通过还原气体将氧化铁原料还原为金属铁。

7. 由于立式竖炉结构简单，可以使用通过了验证且改动最少的现成设计，这也是它吸引人的地方。

8. 新的替代方法试图在一个反应器中完成气化、还原、熔炼和精炼所有四个步骤，将铁矿物炼制为钢。还没有商业案例证明该方法可行。

9. 温度为800℃至900℃的气体还原剂被送入还原炉，与向下的铁矿石逆向而行，从顶部逸出，在那里重新参与还原工艺流程，或输出发电，或输出生产化工产品。

10. 并非每一家工厂都可以利用直接还原的所有优势，因为有多种直接还原工艺可供选择，并且每种直接还原工艺都有其独特的优势。

练习七

1. 在主吹氧期间(吹氧量占 25%—80%)，它性能稳定，熔点为 2573 K。

2. 在 BOF 工艺中，铁水中的 Mn 含量通常在 0.3—0.6wtpct 之间变化，并且发现预计的平衡值比实际锰浓度低 1000~10000 倍。

3. 一氧化碳气泡成核导致膨胀，引起密度不断变化，因此用力平衡法算出液滴在水平和垂直方向上的运行轨迹。

4. 在冷却流程的第三步，将气体还原剂(75%的氢、14%的一氧化碳、7%的二氧化碳、4%的甲烷)从重整器引入固定床反应器。

5. 作为最早的直接还原工艺，第一代和第二代希尔法取得了早期的成功，尽管如此，但它们无论是在能耗方面，还是在运营和投资成本方面，都不再具有竞争力，因此被更成功的第三代希尔法取代。第三代希尔法于 1979 年被研发出来。

6. 该工艺涉及三个主要部件：还原铁矿石的竖式反应器、重整天然气获取一氧化碳和氢的气体重整器以及气体还原剂加热器。

7. 第二次世界大战后，托马斯或者贝塞麦炼钢法不再是氧气底吹，而是氧气顶吹，吹氧炼钢技术就广泛流行起来。

8. 出铁后，炉渣浮在铁水上面，通过"撇渣器"与铁水分离。铁水由撇渣器下部流入铁槽(小井)，然后越过"铁沟坝"进入铁沟。

9. 合金钢分为两种类型：添加元素低于 10%的低合金钢和高于 10%的高合金钢(后者的添加元素一般在 15%—30%之间)。

10. 钨主要和铬一起用作高速工具钢的元素，这种钢的含钨量为 14.00%—18.00%，含铬量为 2.00%—4.00%。

练习八

1. 须知当前的发展趋势是大力研发和应用能够保护自然资源和生态环境的技术，包括降低二氧化碳排放的技术，因此，电弧炉作为基于废钢的钢冶炼综合设施是一种特别吸引人的解决方案。

2. 其他一些须考虑的因素包括：尽量降低可能会损坏电极的废钢的塌陷；确保大型重块废钢不会正好位于燃烧器端口前，因为这将导致火焰反吹到水冷面板上。

3. 熔池中氧气和碳发生反应生成一氧化碳。在炉内氧气充足的情况下，一氧化碳会燃烧，或者通过直接排空系统被排出、燃烧，然后再被输送到污染控制系统。

4. 电弧炉炉料中常有较高浓度的磷和硫，它们通常是钢中不允许存在的，必须去除。

5. 电弧炉熔炼车间的电气系统通常包括一个从电力公司供电的主系统和一个将电力公司输送来的电力电压降低的二次电气系统，二次系统向电弧炉供电。

6. 石墨电极由精细分割的煅烧石油焦组成，混合约 30%的煤焦油沥青作为粘结剂，再加上每个制造商特有的专有添加剂。

7. 历史上，电极的消耗曾经高达 12—14 磅/吨钢，但得益于电极制造和炼钢流程的不断优化，电极的消耗量已降低到约 3.5—4.5 磅/吨钢。

8. 一氧化碳气泡炸裂造成的钢水和渣滴的溅射一直被当做电弧炉灰尘排放的主要原因。

9. 直流和交流电弧炉构成了输变电系统中扰动最为强烈的负荷之一：交、直流电弧炉的特点是吸收功率变化迅速，尤其是在熔炼的初始阶段，在这个阶段，断弧的临界条件可能会变成短路或开路。

10. 今天设计的电弧炉额定输入功率都非常大。由于电弧和熔炼过程的性质，这些电弧炉设备能够引起电网严重的电力质量问题，主要有谐波、间谐波、闪变和电压不平衡。

练习九

1. 提高电机效率对节能具有重要意义。因此，许多研究人员致力于开发和生产高磁感应、低磁芯损耗的高效电机用无取向硅钢。

2. 为了控制冷轧前晶粒尺寸，生产高性能无取向硅钢，对带热带正火的一段冷轧工艺和带中间退火的两段冷轧工艺进行了比较研究。

3. 两段轧制法虽然会稍微增加生产成本，但有利于生产高性能无取向电工钢，在某些特定应用领域具有一定潜力。

4. 晶粒的快速生长是由于在较高温度下晶界迁移率的增加。此外，这种不均匀的微观结构还归因于初始微观结构的继承。

5. 随着热轧温度从 1000℃ 降低到 800℃，均匀等轴的前奥氏体晶粒组织转变为层状和超细前奥氏体晶粒交替排列的非均匀组织。

6. 高强度超重型钢板广泛应用于建筑结构、桥梁、海洋平台等领域。近年来，低碳、中锰变形诱导塑性（TRIP）钢因其优异的高强度和高延性结合性能而受到广泛关注，成为先进厚钢板的良好候选材料。

7. 目前对中锰钢的研究主要集中在化学成分和热处理参数方面，对不同热轧温度下的初始组织研究较少。

8. 层状残余奥氏体的宽度随着轧制温度的降低而增大。结果表明，随着轧制温度的降低，锰在临界退火过程中扩散得更充分。这是因为 Mn 元素沿位错和晶界通道扩散，而 R800-IA 样品由于轧制温度低而位错密度高。

9. 800℃ 时的原子能量低于 900℃ 和 1000℃ 时的原子能量，800℃ 时恢复和再结晶受到抑制。因此，部分前期奥氏体晶粒在轧制后保持拉长形态，在后续直接水淬时变为马氏体。

10. 在这种新型波纹辊轧制工艺中，上波纹辊与变形阻力大的金属 1 接触，下波纹辊与变形阻力小的金属 2 接触。通过这种新的轧制工艺，可以将两种或多种不同的金属加工成上表面和结合面呈波纹状的波纹复合板。

11. 轧机是这一过程中的主要设备,其自动控制是轧钢领域自动化要求最高的部分。轧制力的确定在轧制生产中具有重要意义,可以为轧辊间隙的设置提供依据,指导设备承载力的设计或选择和强度校核。

12. 除了动态轧制力外,该模型还可以很容易地得到特定的轧制压力分布和沿投影接触弧的水平应力分布,这是用平板法建立数学模型的优点之一。

练习十

1. 火法冶金中用到的大多数反应都是可逆的,因此一旦反应物形成产品、达到一种平衡后,预期产品就会迅速转化为反应物。

2. 吉布斯自由能衡量的是让一个反应产生的热力学驱动力的大小。计算吉布斯自由能的方程如下式:

$$\Delta G = \Delta H - T\Delta S$$

式中,ΔH 是反应过程中的焓变,T 为绝对温度,ΔS 为反应过程中的熵变。

3. 上个世纪,人们研发了低温湿法冶金技术。这种技术因为能够处理与硫化铜矿共生的低品位氧化铜矿,所以一直是传统熔炼技术的补充。

4. 虽然铝元素在地球上非常常见,但是一直没有实用的办法以合理的成本生产出铝金属,直到发生了两次技术上的突破:一是找到了从铝土矿生产提纯氧化铝的办法(拜耳法),二是发明了将氧化铝转化为金属铝的方法(霍尔—埃鲁法)。

5. 金红石和钛铁矿都是坚硬致密的矿物,不易风化,因此它们往往富集在砂矿中,通过疏浚开采,并通过密度、磁性和静电等综合分离方法与其他有价值的矿物分离。

6. 分析图 23.2.2 中所示的埃林汉姆图,可以看出为什么古人知道的金属主要是那些可以在相对低温下冶炼就能得到的金属如铜和铅,因为当时已有方法燃烧木炭获得热量和碳实现这种相对较低温度。

7. 目前正在进行进一步的化学和矿物学分析以及质量平衡计算,旨在确定是否使用了任何助熔剂,更具体地说,是否在炉料中添加了铁或石英来帮助形成铁铝酸盐渣。

8. 事实上,在从硫化矿石中获取组分金属的每个过程中,电化学机制都起着关键作用。这些过程包括矿石(硫化物矿体的风化)的勘探、矿石有价组分的分离和浓缩(浮选)、矿石有价金属成分的溶解(萃取冶金)、溶液中金属的回收(电积)和金属的提纯(电解精炼)。

9. 浮选过程不仅可以使有价值的硫化物矿物从组成矿石基质的无价值的脉石矿物中富集出来,而且还可以使不同的硫化物矿物相互分离。

10. 环境对冶炼中含硫气体排放有限制,硫酸的市场有限,这些使得人们努力寻找能从硫化物矿物中提取金属的湿法冶金工艺。

参考文献

1. Christiane Nord. *Translating as a purposeful activity: functionalist approaches explained*[M]. Shanghai: Shanghai Foreign Language Education Press, 2001.

2. Cosmo Di Cecca, et al. Thermal and chemical analysis of massive use of hot briquetted iron inside basic oxygen furnace[J]. *Journal of iron and steel research, International*, 2017(24).

3. 艾星辉. 金属学[M]. 北京: 冶金工业出版社, 2009.

4. 包彩霞. 汉语的定语与英语的定语从句[J]. 北京第二外国语学院学报, 2004(2): 9-14.

5. 陈脑冲. 论"主语"[J]. 外语教学与研究, 1993(4): 1-9, 80.

6. 《词典》编辑组. 英汉冶金工业词典[M]. 北京: 冶金工业出版社, 1999.

7. 方梦之. 中国译学大辞典[M]. 上海: 上海外语教育出版社, 2011.

8. 冯伟年. "外位语"结构在翻译中的运用[J]. 外语教学, 1991(4): 52-56.

9. 贺文照. 论中国传统译论中的读者观照[J]. 外语与外语教学, 2002(6): 44-46, 52.

10. 黄国文. 语篇分析概要[M]. 长沙: 湖南教育出版社, 1988.

11. 黄忠廉. 翻译变体研究[M]. 北京: 中国对外翻译出版公司, 2000.

12. 吕叔湘. 汉语语法分析问题[M]. 北京: 商务印书馆, 1979.

13. 吕英莉. 科技英语的翻译特点与翻译策略研究[J]. 有色金属工程, 2022, 12(7): 195-196.

14. 蒲筱梅. 科技英语定语从句汉译的句法重构[J]. 云南农业大学学报, 2010, 4(04): 91-93.

15. 清华大学《英汉技术词典》编写组. 英汉技术词典[M]. 北京: 国防工业出版社, 1985.

16. 宋春青, 张振春. 地质学基础(第二版)[M]. 北京: 高等教育出版社, 1982.

17. 宋赞, 李相帅, 查春和. 我国直接还原铁工艺的发展现状及趋势[J]. 冶金管理, 2020(16): 22-24.

18. 泰利科特. 世界冶金发展史[M]. 周曾雄, 华觉明, 译. 北京: 科学技术文献出版社, 1985.

19. 童养性. "外位语"之说当用于英语教学[J]. 外语教学, 1983(4): 25-27.

20. 魏寿昆. 选用"碳""炭"的简明方法[J]. 科技术语研究, 2006(3): 4-5.

21. 邢福义, 汪国胜. 现代汉语[M]. 武汉: 华中师范大学出版社, 2006.

22. 徐匡迪. 中国特钢生产60年[J]. 钢铁, 2014, 49(7): 2-7.

23. 徐树德, 赵予生. 英汉钢铁冶金技术详解词典[M]. 北京: 机械工业出版社, 2015.

24. 阳琼. 大数据下的科技术语音译[J]. 中国科技翻译, 2018, 31(3): 1-5.

25. 杨枕旦. stratosphere, 从"同温层"到"平流层"——科技术语翻译杂议(七)[J]. 外语教学与研究, 2000(6): 461.

26. 张立峰. 炼钢技术的发展历程和未来展望(Ⅰ)——炼钢技术的发展历程[J]. 钢铁, 2022, 57(12): 1-12.

27. 周兰花. 冶金原理[M]. 重庆: 重庆大学出版社, 2016.

28. 周晓梅. 中国文学外译中的读者意识问题[J]. 小说评论, 2018(3): 121-128.

29. 朱徽. 汉英翻译教程[M]. 重庆: 重庆大学出版社, 2004.